Fighting for Birds

25 years in nature conservation

Mark Avery

Published by Pelagic Publishing
www.pelagicpublishing.com

Fighting for Birds: 25 years in nature conservation

Published by Pelagic Publishing
www.pelagicpublishing.com
PO Box 725, Exeter, EX1 9QU, UK

ISBN 978-1-907807-29-9 (Pbk)
ISBN 978-1-907807-31-2 (Hbk)
ISBN 978-1-907807-30-5 (eBook)

British Library Cataloguing in Publication Data
A catalogue record for this book is available from the British Library.

Foreword

I started this book at the beginning and read to the end without stopping and that, for me at least, is a rarity. But then Mark Avery is a rarity too, as he has few, if any, equals in contemporary British conservation and I was absolutely itching to discover his ideas, views and visions in this eagerly awaited book. It's been a long time in the making; Mark is in his fifties, but it's been worth the wait and there is no doubt that this book will be recognised as a truly important one. I don't use this term glibly either, as I can only think of one other in this category that has made it onto my shelves in recent years. But for 'important' please do not read in any way 'worthy', it's just that here a tremendously respected figure has confronted a series of critical issues with dynamic aplomb. And I greatly enjoyed it too. I didn't agree with all of it of course, but if anyone has the passion and reason to change minds on all matters wildlife and conservation it is Mark Avery. So I learned a tremendous amount, understand some key issues far better now, sniggered at some very tasty gossip and admired the pragmatism and clear thinking that has formulated his authoritative views. It made me think and that is always an essential component to a good read.

It all begins so gently with typical boyhood tales of a budding birder, before describing a formative education in the company of some of recent ecology's true giants and some fascinating science, all enthusiastically explained, and then charts Mark's years of influence and success at the RSPB. But this is very much a platform for what follows: honest behind the scenes details and clear explanations of the recent evolution of this influential charity, all with a fair spattering of candid criticism thrown in. It therefore addresses many of the criticisms and concerns which unsettle some of the membership, and I'm sure will lay most to rest. Ultimately however, it is the climax of the book which holds most reward because here one of our most respected and knowledgeable conservationists gets straight and serious on the issues which command so much of our concern. And I'm sure some will find controversy here. Not me. I read and saw real clarity, and I felt heartened and re-assured that there are those with the courage to tell the hard truth about the health of, and future hopes for, our nation's wildlife. So for myself it's a triumph, and if you have any real interest in the job of saving species and their habitats then it's a tremendously rewarding 'must read'.

Chris Packham, April 2012

P.S. I can't wait for volume two… maybe he will deal with domestic cats!

Preface

Writers write to influence their readers, their preachers, their auditors, but always, at bottom, to be more themselves.
Aldous Huxley

This is a book about birds and wildlife and how to save them. It's told through the perspective of my 54 years on this planet, during most of which time I have been fascinated by, and in love with, the natural world.

If you were to be utterly conventional and start at the beginning of the book and read through to the end, then you would progress from the past to the future, and from the personal to the general. Throughout, I try to extract particular lessons and general truths from the instances and events described, whilst also trying to give the smells and the flavours of work in conservation – as it is actually done.

There is a lot of me in this book, but it's certainly not an autobiography. There is nothing here of girls, marriage, family or love – except a love affair with the natural world. But in writing about nature I have dipped into my own experiences and work over the years to tell a story of how nature conservation works in the UK. So there is quite a lot about birds I saw, places I visited and people I met – but the hero, or heroine, is definitely Nature, not me.

The first chapter skips through more than half my life to explain the influences that made me a nature-lover, birder, naturalist and conservationist, and ends with me arriving at the RSPB in 1986 as a scientist. The next two chapters describe some of my early work as a scientist at the RSPB involved in upland afforestation issues in the far north of Scotland and roseate tern conservation from the UK to West Africa. These two areas, between them, introduced me to a wide range of the RSPB's conservation work from policy to nature reserves in the UK and beyond.

The bulk of the book consists of stories about different aspects of nature conservation told through my experiences at the RSPB but also trying to extract the bigger messages from these events and to develop an intellectual framework for nature conservation. This book is not a day-by-day, month-by-month, nor even year-by-year account of the life and times of an RSPB person. Rather, it is a collection of thoughts and reflections on the birds, places and people that I've encountered over a 25 year career in the world's best nature conservation organisation – more than half of that time spent in a senior post as Conservation Director.

The last few chapters try to make some sense of the broader state of UK nature conservation with its tangled bank of wildlife NGOs, and sets out some challenges for all of us who want to make the world a better place for nature.

The views expressed in this book are mine, not those of the RSPB. Some of them are the views that I had to keep to myself as an employee and am now free to voice as a freelance writer and environmental commentator (Chapters 10, 11, 15 and 16 are the main places to find these).

If you would like to keep in touch with my writing and thoughts then I write a daily blog about UK wildlife issues at www.markavery.info and a monthly column, 'the political birder', for *Birdwatch* magazine.

Mark Avery, Northants, March 2012

> *Writing a book is an adventure. To begin with, it is a toy and an amusement; then it becomes a mistress, and then it becomes a master, and then a tyrant. The last phase is that just as you are about to be reconciled to your servitude, you kill the monster, and fling him out to the public.*
> **Winston Churchill**

Acknowledgements

I'd like to thank my family members (parents, wife and children) for putting up with me while I lived the experiences described in this book (and those that aren't) and then for putting up with me again as I wrote about those experiences.

I don't think that I would have ended up having so much fun in nature conservation if it weren't for the two school masters mentioned in Chapter 1, the late Derek Lucas and Tony Warren.

I wouldn't have had much of a career as a scientist if I hadn't got off to a reasonable start with the help of Professor Tim Clutton-Brock FRS, Professor Lord Krebs FRS and Professor Paul Racey. Although they all sound terribly grand and important, and indeed they are, they were all kind and helpful to me at times when I needed kindness and help, and I thank them all for the chances they gave me.

The late Colin Bibby recruited me to the RSPB and had popped in and out of my life in the years before. He was a great intellectual birder and influenced so many of us so deeply.

When Colin moved on from the RSPB, Graham Wynne made the brave decision to appoint me as Colin's replacement as Head of Conservation Science, and then several years later, when Graham became RSPB Chief Executive, he made another brave decision to appoint me as Conservation Director. I worked directly to Graham (now Sir Graham) for about 19 years and I can honestly say that he was only completely unreasonable a few times each year. I owe Graham an awful lot, as do so many others, and as does UK nature conservation, and it was a privilege to work closely with him for so long.

The RSPB was a friendly and inspiring place to work. All my 25 years at the RSPB were happy ones and very few of the days were unhappy – that's quite a thing to be able to say. Thank you to all my ex-colleagues from the lads in the Transport Office to those who worked closely with me, and from my Board colleagues to my long-suffering PAs (Anita McClune and Claire Farrar).

I'd like to thank my successor at the RSPB, Martin Harper, for reading through this book on the RSPB's behalf and not being too picky about it!

Thank you also for pointing out one lapse of my memory which I have corrected but I wonder how many remain. Good luck in a great job and just remember not to be too nice to too many people!

And there was this bloke I met… No, that's enough thanks. There are so many people in nature conservation who are doing a really good job under difficult circumstances because they care about nature. To those I've worked with – thank you. To those I have yet to meet – I'm looking forward to it! Let's hope we all win in the end.

Contents

List of abbreviations

ADAS	Agricultural Development and Advisory Service, 1971–1997 and continuing as a private concern, *ADAS*. Originally 1946–1971 as the National Agricultural Advisory Service (NAAS)
ASSI	Area of Special Scientific Interest
BBS	Breeding Bird Survey
BOU	British Ornithologists' Union, 1858–
BTO	British Trust for Ornithology, 1932–
CAP	Common Agricultural Policy
CBC	Common Birds Census
CCW	Countryside Council for Wales, 1990–*
CROW	The Countryside and Rights of Way Act of 2000
DECC	Department Energy and Climate Change, 2008–
Defra	Department for Environment, Food and Rural Affairs, 2001–present**
DoE	Department of the Environment, 1970–1997**
EA	Environment Agency, 1996–**
EC/EU	European Community, 1967–1993 / European Union, 1993 – present. Previously the European Economic Community 1957–1967)
EN	English Nature, 1990–2006*
ESA	Environmentally Sensitive Area
FBI	The Farmland Bird Index
FC	Forestry Commission, 1919–
FoE	Friends of the Earth, 1969–
GCT/GWCT	Game Conservancy Trust, 1931–2007 / Game and Wildlife Conservation Trust, 2007–
IUCN	International Union for the Conservation of Nature and Natural Resources, 1956 – present. Previously the International Union for the Preservation of Nature (IUPN)
JNCC	Joint Nature Conservation Committee, 1990 –*

NCC	Nature Conservancy Council, 1973–1990*
NE	Natural England, 2006–*
NERC	Natural Environment Research Council. An umbrella organisation with various responsibilities, 1967–
NFU	National Farmers Union, 1908–
NGO	Non-governmental Organisation (a generic term)
NIEA	Northern Ireland Environment Agency, 2008–*
Ramsar	The Convention on Wetlands of International Importance, 1971. Not actually an abbreviation, but the name of the place in Iran where the participating countries signed the convention, which names important wetland sites, called Ramsar sites.
RBBP	Rare Breeding Birds Panel
RSPB	Royal Society for the Protection of Birds, 1891– (Royal Charter awarded 1904)
RSPCA	Royal Society for the Prevention of Cruelty to Animals, 1824–
SAC	Special Area of Conservation
SCARABBS	Statutory Conservation Agencies and RSPB Annual Breeding Bird Surveys
SCI	Site of Community Importance
SEO	Sociedad Espanola de Ornithologia, 1954–
SNH	Scottish Natural Heritage, 1990–*
SPA	Special Protection Area
SSSI	Site of Special Scientific Interest
WeBS	Wetland Bird Survey
WWT	Wildfowl and Wetlands Trust, 1946–. Initially the Severn Wildfowl Trust and then the Wildfowl Trust.

*** Key dates in the history of UK Agencies responsible for wildlife matters**

1949	The Nature Conservancy (NC) established by act of parliament.
1973	Nature Conservancy becomes The Nature Conservancy Council (NCC).
1990	NCC split into four parts: English Nature (EN) / Scottish Natural Heritage (SNH - formed from Scottish part of NCC and the Countryside Commission for Scotland) / Countryside Council for Wales (CCW – similarity formed) / and the Joint

	Nature Conservation Committee (JNCC – to serve all three agencies).
1990	Countryside Agency: formed to take on the English responsibilities of the Countryside Commission (previously the Countryside Commission for England and Wales, then the Countryside Commission for England).
2006	Natural England (NE) formed from English Nature and parts of the Rural Development Service and the Countryside Agency
2008	Environment and Heritage Service renamed the Northern Ireland Environment Agency (NIEA).

**** Key dates in the development of the Department of the Environment**

1970–1997	Department of the Environment (DoE).
1997–2001	Department of the Environment, Transport and the Regions (DETR).
1996–	Environment Agency. Formed to take over from the National Rivers Authority, Her Majesty's Inspectorate of Pollution, and the waste regulation authorities in England and Wales. Currently part of Defra.
2001–	Department for Environment, Food and Rural Affairs (Defra).

Chapter 1

Early years

Bird on the horizon sitting on the fence
He's singing his song for me at his own expense
And I'm just like that bird oh oh
Singing just for you
Bob Dylan

I was born quite early on the Saturday morning of Grand National day, 1958. Charles Avery, my father, telephoned my mother Megan's relations, the side of the family living in south Wales, to tell them the news. In those days few had telephones so the call went from my father to my uncle Peter who then spread the word of Megan's child through the terraced houses of the Pontypool mining community.

The message was passed concerning my arrival, gender, weight, number of fingers and toes, blue eyes and the general well-being of mother and baby. Uncle Peter then passed on the news through the day but soon realised that he and my father had dropped the ball somewhere in their conversation – what was this new baby's name? After quite a bit of censure and teasing from the female side of the family (as he told me over 40 years later) he made his annual flutter on the big race. Scanning down the list of runners there was one that caught his eye because of the morning's events, and he backed the winner of the Aintree National, *Mr What*, at 18/1 thanks to my birth.

Despite the fact that National Hunt racing has been a lifelong interest, that is almost the last you will hear of it in this book which is not an autobiography. Rather, it is an account of experiences and thoughts about the world of nature and nature conservation. This chapter takes you from the day of the 1958 Grand National through to 1 April 1986, when I joined the staff of the Royal Society for the Protection of Birds (RSPB), and is a quick canter through the events that led to me working for the UK's best nature conservation organisation.

This chapter shows how random events – like a chance meeting in a pub – gave my life the nudge that sent it forward in a particular direction. It wasn't inevitable that 28 years after *Mr What* won the Grand National I would join the RSPB and work there for the next 25 years, but it was those apparently random choices and chances of friends and events that, looking back, and only looking back, made an RSPB career inevitable and just what I needed in life.

Early influences

All my early years were spent either in Bristol or nearby. My father was a Bristolian whereas my mother was a miner's daughter and nurse from south Wales. We first lived in the southern suburb of Brislington, close to where my father had been born, and then moved to almost within sight of fields in Whitchurch, and then further south again into the north Somerset village of Pensford, just seven miles from Bristol city centre but out in the countryside.

Our holidays were spent in places like the Lake District, New Forest or mid-Wales and Sunday afternoons usually included a drive in the countryside whether to the Mendips, the Somerset Levels or the Cotswolds.

I remember Dad pointing out the larger and commoner species such as kestrels, green woodpeckers and buzzards and we always thrilled when we saw a fox or a deer. But the countryside and the nature which lived in it were not in any way thrust upon me. They were on offer but no more so, as I can recall, than church architecture, cars, books, music, sport or a host of other interests. I collected stamps, made Airfix model aeroplanes, played war games and read voraciously from Enid Blyton through Conan Doyle to much of Hardy and most of Dickens, all by the age of 11.

You'll note that these were solitary pursuits. I was an only child and throughout my early education went to schools that were not those frequented by my neighbourhood friends, so was not surrounded by school friends at home. Instead of the local primary school my parents sent me to a small private school, Cleve House School, a short bus ride away, because they were worried that my youthful stammer would not be treated kindly by the fierce head teacher at the local primary. Cleve House gently slowed the pace of my speech so that my lips and brain were in synch, and gave me a good enough education to pass the exam to go to the Bristol Cathedral School a year early. But it was decided that I would take the entrance exam again the next year and stay at Cleve House rather than be a year younger than my new classmates.

We had moved from Whitchurch to Pensford by the time I retook and passed the entrance exam the next year, and had switched our target to the Bristol Grammar School – the largest of the three boys' direct grant grammar schools in Bristol. The move to Somerset at the same time meant that even though my exam results would have won me a free place at the Cathedral School, because we now lived in Somerset (rather than in the City and County of Bristol or Gloucestershire) my parents had to pay for my Grammar School education. I was the last generation of school children to take the 11-plus exam, and I passed, so the Cathedral School at Wells would have been another option but the Comprehensive system arrived and it was either the local Chew Valley Comprehensive or Bristol Grammar School for me. I'm so grateful that my parents, who were by no means well off, made the sacrifices to send me to Bristol Grammar School – it not only gave me an education that set me up for the rest of my life, but gave me the opportunity to make birds and nature a career too.

Bristol Grammar School

I made the daily journey to school, wearing my school uniform of a grey suit, school (or House) tie and cap (up until the age of 14, I think) on the 376 or 377 bus from Pensford, up the A37 Wells Road into Bristol Bus Station after which there was a calf-stretching uphill walk to the Grammar School's Victorian buildings which stood next to many of Bristol University's buildings at the top of Park Street.

Making its way through the rush hour traffic, the bus would be full of office-workers and a very few other school children (including some very fetching miniskirts from the Catholic girls grammar, La Retraite). I remember the bus windows being steamed up on winter's days as I did the remains of my homework (maybe Latin or physics) on the journey. On the return journey in the winter evenings I always looked up at the massive swarm of starlings circling above Brunel's Temple Meads Station.

The school week at Bristol Grammar consisted of three half days (Tuesday, Thursday and Saturday mornings) and three full days (Mondays, Wednesdays and Fridays) with at least one of the half days also being a sports afternoon. So I would be playing rugby or cricket on a Tuesday or Thursday, and just occasionally on a Saturday too.

Sundays were free except that every other Sunday during term two masters, Derek Lucas and Tony Warren, would take a minibus full of teenage boys out into the Somerset or Gloucestershire countryside on a Field Club outing.

When I arrived at the grammar school as a callow and nervous 11 year old I looked through the list of clubs and societies and the only one that caught my eye was the Field Club. This was the school natural history society which had indoor meetings as well as the weekend excursions. I wanted to be a member and was aghast to discover that entry was only open to boys in their fourth year or above (Lower Removes or Removes – don't ask!) rather than first-years (Third Formers – I said, don't ask!) like myself. A group of us, Andrew Brown, Peter Davies and Ian Cree (from memory) all discovered each other and our shared desire to break into this select society in those first few days and persuaded Mr Lucas and Mr Warren to relax the age constraint and allow us membership. From then on, the Field Club was as important a part of my secondary education as were O and A Levels and Cambridge entrance examinations. When I hear talk of a grammar school education I think of A Levels, the Golden Hill rugby pitches and learning to tell bar-tailed and black-tailed godwits apart at Stert Point in Bridgwater Bay.

Over the years the friends I made in the Field Club were longer lasting, and more memorable, than those I made on the sports field or in the classroom. Steve Albon was a few years ahead of me and went on to be a joint editor of the UK National Ecosystem Assessment, after working at the Institute for Terrestrial Ecology and the MacAuley Institute. Hugh Brazier edited *Irish Birds* when based in Dublin. Peter Dolton now checks the golden orioles at Lakenheath RSPB nature reserve. My exact contemporary, Peter Fraser, is a joint author of a Poyser monograph and Tim Dee, who came up through the ranks a few years after me, is the author of a marvellous book about nature. And there were many others who made those Field Club meetings and trips a delight.

Imagine about a dozen teenage boys all piling out of a minibus with binoculars, notebooks, spots, telescopes, waxed jackets, wellies and masses of testosterone. If you didn't learn to identify birds in that company you were toast! Don't get me wrong, it was a friendly crowd, but a competitive one too – we were grammar school boys after all. If you identified a difficult or distant bird correctly, or spotted a good bird first, then your reputation for the day was made, all the more so if you were a youngster and were praised by the older boys, of course. But get it wrong, and you wouldn't be allowed to forget it for that day at least. Peer pressure is a great thing to encourage improvement. I started as a novice but rapidly developed into a pro (for my age, at least).

Our visits to local sites marked out the seasons: September and October was for high-tide visits to Bridgwater Bay; mid-winter took us to Slimbridge for geese (hopefully including a lesser white-front) and the Somerset Levels for wildfowl and waders; while spring involved a trip to Brean Down for arriving migrants such as grasshopper warblers and an evening trip to Shapwick Heath for nightjars and nightingales. Throw in a few visits to the local reservoirs of Chew Valley and Blagdon, some dull woodlands and Sand Point or Avonmouth, and the year was more or less full. As time went on some longer trips were arranged too – to Exmoor and Dunkery Beacon, the boat trip to Lundy from Weston-super-Mare, a weekend to Dumfries and Galloway and trips to the Exe Estuary, the Berkshire Downs and Portland Bill.

Those trips taught me a lot about birds and a lot about being a teenage boy. They were bonding experiences and although some must have been rain-soaked and birdless, I remember the highlights not the low points. On what may have been my first trip we saw a purple sandpiper sitting on the rocks at Sand Point on the Severn Estuary, north of Weston-super-Mare, that was so close that I didn't see it for ages because I was looking too far away. It looked just like its picture in the field guide (the Peterson, Mountford and Hollom one) with its dumpy dark body and yellow legs. I didn't know much more about it and I certainly didn't know that less than a decade later I would be searching for purple sandpiper nests on the Hardangervidda plateau in southern Norway.

Sparrowhawks were unusual in those early 1970s' days and seeing one, particularly seeing one well, made the day notable by itself. On the other hand, farmland birds were pretty common and I clearly remember, at least it seems clear to me, a flock of a thousand linnets by the side of the road near Priddy, on the top of the Mendips, with a single greenfinch among them. The Somerset Levels in winter were alive with waders – crossing a splashy field on those flat peatlands might flush a hundred snipe or more with the chance of a Jack snipe amongst them and the lapwing and golden plover numbers were huge. And we saw some decent rarities too – sometimes we did find the lesser white-front amongst the thousands of white-fronted geese at Slimbridge and there were occasional pectoral or buff-breasted sandpipers on the coast or at the reservoirs in autumn.

Mostly though, it was a process of gradual accumulation of knowledge of birds from different habitats and at different times of year. It was a time before pagers and long before the internet. The first breeding atlas was published in 1976 – the year after I left BGS and I learned my birds mostly by seeing them in the field on those Sunday trips. I could not have got to all those places myself and nor could I have seen all those birds or gradually got to know what they looked like and sounded like without those Sunday Field Club trips.

Derek Lucas (who was my form master one year, taught me English another year, and was my House Master for many years, and so taught me rugby as well) and Tony Warren (who taught me Geography for a year or two) were lovely men. My companions and I owed them a lot for the opportunities they gave us all. They were civilised and cultured with a love of literature and opera and a remarkable tolerance of teenage boys growing into young men. I hardly remember them ever pointing out a bird and it would be pretty unusual for either of them to identify a bird ahead of the eager younger hordes. In fact, their interest in birds seemed pretty low-key. I wonder what they themselves got out of it all but I do know that they gave me and others a priceless opportunity to learn about birds. I am sure we thanked them politely but now looking back I wonder whether they knew quite how much they affected the lives of many of us.

And it is difficult to imagine two male teachers feeling able to devote this type of time to a bunch of pubescent schoolboys these days. They might neither dare nor be allowed. And that is a shame. Those days of local birding, in a friendly group, under the protection of two adults but with the freedom to be out and about away from parents (and to sneak a crafty first (and almost last) cigarette at the back of the group) were precious days of finding oneself and growing up, as well as discovering the diagnostic differences between marsh and willow tit and adding to one's life list. I would not have developed a strong love of birds, which spilled over into passion for the whole of nature as life went on, without those early Field Club experiences. Without them I cannot see that I would have ended up with the career I've had and I wonder where the next few RSPB Conservation Directors are discovering their knowledge of and love for birds and nature?

Pensford

But there were holidays too and weekends without sport or Field Club trips. I spent those times walking through the local fields along the River Chew.

My favourite walk was down to Publow where the church formed part of the dowry of Anne Boleyn and where an ancient bridge crossed the river. Standing on the bridge at different times of year I saw my first redpolls in the riverside alders and watched a fisherman's line get unbelievably tangled as he pulled an eel out of the river. Those same alders once held my first lesser-spotted woodpecker, which was then a regular sight in that spot and along the river. Water voles were common and the sound of their 'plop' as they dropped into the water from the bank, or the sight of them making a V as they swam across the river was an encounter made at every visit. One day a strange bird was sitting on a branch over the river – grey and upright, it darted out after insects and returned to the same perch time after time. I didn't know what it was until I returned home and found it was a spotted flycatcher. This gradual unfolding of nature's secrets was the prize to be gained from a regular walk. And remarkably all those species I have mentioned are now so much less common. No-one would have guessed their fate back then.

The spring would be heralded by the bright buttery, male brimstone butterflies from March (or sometimes February) onwards – flying through the woods when there were still rather few leaves on the trees. Later the wood whites would fly the same way but by then the trees were fully green. The species that would always spark my parents' interest was the kingfisher, which I sometimes used to see perched and watch while it fished. But more usually, it would see me first as I rounded a corner of the river and my view was of its turquoise back heading rapidly away from me along the river course. There were human fishermen too, and just occasionally I would pass Pensford-born Acker Bilk sitting by the riverside. I cannot hear his *Stranger on the Shore* clarinet melody without being taken back to a slow-moving wildlife-rich river with its water voles and kingfishers.

Our house backed on to farmland and a 20-mile view east to the hills above Bath, and the knot of trees by Bath racecourse. While doing my homework I used to gaze out over that view and occasionally see a distant buzzard. Our garden fence was almost as likely to have a tree sparrow on it as a house sparrow and in spring we were occasionally woken by a singing cuckoo on the garden fence. I made a nest box in woodwork lessons which was later occupied by blue tits, and a pair of goldfinches nested in the American currant bush outside the kitchen window.

One winter there was a massive cold weather movement of lapwings, golden plover, a few snipe and ducks that passed over our house heading west to the coast and warm West Country from the frozen east. I sat and watched and counted as flock after flock of lapwings passed overhead. I'd never seen anything like it before and I've never seen anything quite like it since. And one spring a corncrake spent a night or two singing in the field at the bottom of the garden before heading north for the Hebrides.

Once I reached the age of about 13, I was able to set off on my bike to more distant parts. This would sometimes be a ride around the local lanes just to see what was happening but mostly it was a trip to Chew Valley Lake, a distance of about five miles through the lanes of north Somerset. I used to meet friends there sometimes but I was quite happy spending a day alone, cycling around the lake, calling in at the hides and just seeing what birds were around.

The cycle to Chew Valley Lake was always full of hope. Even if it were another visit following three previous days when there was little about, there was always the chance of a rare bird or an amazing sighting of a common one. But the cycle home passed in inverse relation to the success of the day. If I had had a good day then whatever the weather I would fly up the hills, it seemed to me, as well as down their other sides whereas if the day had been dull the upgrades were steeper and longer and the wind was always against me.

The freedom to head off, without a mobile phone of course, for the whole day and to go wherever I wanted was wonderful. And again I can remember the highpoints far better than the low ones. There were plenty of ducks with smew in the winter and spring garganey being the most prized. At passage wader times, if there was enough mud, then one could clock up a number of interesting species with the chance of a rarity thrown in. Late August was the time to check the black terns for the occasional white-winged black. In winter you might see a Slavonian grebe. Just spending a lot of time there meant that I saw quite a lot of species such as avocet, common scoter, hawfinch and osprey that were passing through since I just happened to be there when they did.

But you just had to take whatever Chew was offering. I remember watching flocks of goldeneyes in late April, with the males throwing their heads back ecstatically in display before they headed north to Scandinavia or maybe to nest boxes on Speyside. One spring a great northern diver stayed right through into late April and I remember hearing it sing early on a misty morning. I've not heard that eerie and arresting sound in the UK since – only from a tent in a wood in Ontario whilst studying bats.

On arrival at the east shore of the lake, I could circle it in either direction but at the far side there was another decision to be made – Blagdon or not Blagdon. Blagdon was another lake, covered by my Bristol Waterworks pass, and another couple of miles further on. Nothing in a car – and delightfully downhill most of the way by bike. But what goes down must come up, so a long uphill return. And it was usually only if the lake had been a bit dull that I even considered Blagdon, unless word was out that there was a good bird there or that there was plenty of mud for waders or some such hint of promise. But I do know that on 14 November 1973 I went to Blagdon. It was a day off from school and while 500 million people were watching Princess Anne getting married to Captain Mark Phillips, I was on my bike in north Somerset. And it was worth the journey as Blagdon produced not only a red-breasted merganser, always a good bird inland in Somerset, but also a great grey shrike which was a very good bird. Such were the simple surprises that gave me pleasure, made the cycling worth it and brought me back time after time during my school days.

As a schoolboy the grapevine wasn't very active. I hardly ever heard about what birds were around until I was very close to them and met a birder who would say what he had seen a few moments ago. So I used to enjoy poring over the monthly newsletter of the Bristol Ornithological Club to see how many records would have my initials next to them and what I had missed at Chew Valley Lake or elsewhere.

There were some initials that used to dominate the interesting records. Brian Rabbits, a shift-working policeman, used to find a lot of good birds at Brean Down and Cheddar when he wasn't working but everyone else was. The initials KEV and TRC often cropped up at Chew Valley Lake to show what Keith Vinicombe and Tim Cleeves had spotted. I made sure that the initials BGSFC became well known and there were a smattering of MIAs too.

The same monthly newsletter sometimes carried trip reports from far-away East Anglia where Cley and Minsmere seemed like good places to go, or from the Highlands of Scotland and increasingly from the Camargue, Spain and even Israel. With my copy of Peterson, Mountford and Hollom I could learn the identification features of birds I might well see and also dream about distant birds that I never really imagined being able to get to. I would spend my pocket money on New Naturalists such as Ian Newton's book *Finches*, or Poyser monographs such as the one by Malcolm Ogilvie on ducks. And John

Gooders' *Where to Watch Birds* was a wonderful guide to places which I could recommend to my parents for our next family holiday.

Reserves of delight

Back in those days, applying to Oxford or Cambridge involved one in taking their individual entrance exams in the term after A Levels and being interviewed at the university in question as well.

So in the last few days of 1975, while *Bohemian Rhapsody* was dominating the charts, I was spending my last days at school and had the first nine months of 1976 'free' before Cambridge University expected to see me at their doors of learning.

I spent February and March 1976 volunteering at Minsmere for the RSPB. The previous summer I had volunteered at Arne for two weeks where I remember the warden Brian Pickess handing me the envelope with my rather good A and S Level results in it. My companion in the basic Arne visitor accommodation was Andy Gosler who has worked much of the time since at the Edward Grey Institute at Oxford, where I would spend some later years too. But, and no disrespect to Andy, sharing the accommodation with him was less exciting than sharing it with the smooth snake that lived under the hut!

Out on the heath there were sand lizards and adders as well as Dartford warblers with coloured rings on their legs. Andy was a ringer and while I knew nothing much about ringing yet, I did note very carefully a certain warbler's ring combination, which leg was which, and which leg carried the standard BTO metal ring. Andy helpfully told me that practically all birds had their metal BTO ring on their left leg, as this one had done, because that was just standard practice for right-handed ringers. I thanked him but did point out that if one varied the location of the metal ring one could have twice as many ring combinations! The Dartfords had been ringed by Colin Bibby, working for the RSPB, and I had already met Colin a couple of times through the Field Club and Derek Lucas. One day Colin came by and asked whether we'd recorded any colour-ringed birds so Andy, as the ringing expert, told Colin about my sighting without mentioning the metal ring. Colin immediately asked which leg the metal ring was on as that, obviously, allowed twice as many combinations – sweet!

But now I was heading for an eight-week stint at Minsmere in winter. No

Dartford warblers there (although there are now) but bitterns, marsh harriers, bearded tits, avocets and so much more. There was some back-breaking work (cutting reeds by hand) and some dull stuff, but lots of bird watching thrown in. There were shorelarks on the beach, hen harriers roosting near the Island Mere hide, bitterns booming and the return of the first spring migrants, including the avocets to the famous Scrape.

Jeremy Sorensen was the relatively new RSPB warden at Minsmere and had moved from the Ouse Washes to take over from the legendary Herbert Axell. Bert had constructed the Scrape and its hides and had made Minsmere famous. Some years earlier I had been on the beach early one August morning looking to rediscover the wryneck, the first I had ever seen, that I had found the morning before. At the Sluice bushes I saw a group of men with mistnets and being a shy teenager I looked on from a distance. I recognised the great Bert Axell as he directed the operations of his staff and volunteers at a mistnet. After a while he beckoned me to come over and thrust a bird under my nose, legs up in his hand. 'What's that then, Sonny?' was the gist of what he said and I replied, 'That's a lesser whitethroat Mr Axell – see, it's got blue legs but looks like a whitethroat with a darker head'. This was the right answer and the eminent Mr Axell replied 'You know more about birds than most of my staff do. The reserve is shut today but you can come in and look around!'. This was an amazing treat for me, as a 13-year old, and Bert and some of his staff spent an hour or so pointing birds out to me before giving me the run of the reserve.

But Jeremy was the new guard and Bert was now living down the road. There was, I gathered, a little tenseness in the relationship and late in my stay Jeremy asked me to go down to the wet woodland on the road to East Bridge and keep an eye, and more particularly an ear, open for willow tits, which Bert apparently swore were present there although Jeremy had never recorded them. I can still remember Jeremy imitating the willow tit's 'Sneh sneh sneh' call as he sent me on my way. I found no willow tits and their disappearance around that time was perhaps one of the early losses of the species, which is now difficult to see over much of southern England.

Jeremy was a short man with a large moustache, and a very rapid walk which I think he had learned in National Service. He was difficult to keep up with. I remember him telling me that the Bewick's swan was probably, in global terms, one of the rarest species I would see in my life.

Also volunteering at Minsmere was a guy called Les Street. Les was a long-term volunteer and soon after I left, his plan worked and he was taken onto the RSPB staff and was to work for them for many years in nature reserves from Islay to the Isle of Wight. He was just starting off then and he and I used to walk the couple of miles or so to the excellent Eels Foot Inn (named after a cobblers' tool) and drink good beer, play bad darts, share a jar of cockles and talk of birds, birdwatchers and nature conservation. Les was a kind man to a young enthusiast. He told me that he had worked in London, I think as a designer of some sort, and had done the daily two-way commute until one day someone had pushed past him onto a tube train and Les had snapped and lifted the man off the tube by his lapels. He realised, for he is a very mild man, that he should find another way to live his life and opted for the relative tranquillity of nature reserves.

I loved Minsmere and it is still one of my favourite nature reserves. When Mum and Dad came to pick me up it was my 18th birthday and my present was a pair of Zeiss 10×40B binoculars – the same pair that I have today and through which I have seen about a 1,000 more species than I had on that day in 1976.

I had just enough time to get my clothes washed before heading off to another nature reserve where this time I would be paid, in charge and on my own. I spent the summer of 1976, one of the driest and warmest summers on record, as the warden of St Cyrus Nature Reserve on the east coast of Scotland in what was then Angus, just where the river North Esk enters the sea. I arrived on the 19 April and stayed through to September. My employer was the Nature Conservancy and I was paid what seemed the enormous sum of £25 a week with free accommodation in the caravan on site. In return I had to protect the colony of little terns nesting on the beach from disturbance.

My school friend Peter Dolton had done this job the previous year and he had tipped me off about it and put in a good word. What a lovely summer – guarding terns from couples trying to find a secluded place to copulate, the occasional dog walker and wandering fishermen. It was a lovely place to spend a summer with a pair of whinchats nesting by the caravan, Maiden pinks in the meadow nearby, and the occasional salmon splashing in the shallows that I could liberate as another perk of the job.

There were also some rare birds – a May honey buzzard, a July little auk (most little auks I have seen have been in odd places at odd times) and a roseate tern one evening.

Each Friday I would walk up to the St Cyrus village shop to cash my wages cheque, pay most of it into my Post Office account, do some shopping, buy a pub lunch and a pint and walk back down through *my* reserve to *my* caravan having seen *my* birds. Even as an Englishman in Scotland, and even as a student summer jobber, the feeling of possession of this place was strong. I could see why Bert Axell behaved as though Minsmere were his. He had much more right to that feeling than I did but I felt it a bit too and later, when working for the RSPB, I have always taken care to consider the understandably strong views of the person on the ground.

Cambridge

In October 1976, after a summer of sun and drought, the government appointed Dennis Howell as minister of drought and the skies duly emptied for the first month I was at Cambridge. I was ensconced in Downing College amongst lots of undergraduates studying Law, Engineering and Medicine – and, of course, all men (or boys, at least).

There were three elements of my time at Cambridge which are relevant to my conservation career – the Cambridge Bird Club, an expedition to Norway and working on Rhum.

The Cambridge Bird Club has an illustrious history and in 1976 when I went up to Cambridge its Chairman was the RSPB Director Peter Conder. During my time at Cambridge various fellow undergraduates were leading lights too – Hugh Brazier (who had left BGS a couple of years before me), Tim Stowe (who left Cambridge for the RSPB and has worked there ever since [now as International Director]) and Simon Stuart (who worked for the International Council for Bird Preservation and then for the IUCN Species Survival Commission).

But it was a chance meeting in a pub (gosh – of all places!) which had a big impact on my future direction. The chance meeting was with ex-BGS friend Steve Albon who asked me what I was doing in the summer vacation. I had to confess that I didn't know, so he suggested that I should apply to work on Rhum on the Red Deer Project there. And so I applied and was interviewed by the chief researcher, Tim Clutton-Brock, now a Fellow of the Royal Society but then an unmarried Kings College Fellow wearing jeans and inhabiting the most amazing rooms in the Gibbs building.

The Gibbs building is the large block to the right of Kings College's famous

Chapel as you gaze at the Chapel from the Backs. In the middle at the top of the Gibbs building is a large semi-circular window and behind that window was where I mumbled answers to a few clipped questions from Tim and somehow persuaded him that I could do fieldwork on wild red deer on the island of Rhum that summer.

The red deer project was based at the far north corner of Rhum, away from the island's few other inhabitants, and our task was to follow individual deer around for a 24-hour period and record what they were doing every minute. Long before the days of Mp3 players we had an earpiece in one ear which bleeped every 60 seconds to signal that we should record what a particular deer was doing. My partner in this work was another Tim, Tim Johnson, who was reading Zoology at Oxford (boo, hiss!) and now works for the World Conservation Monitoring Centre in Cambridge. Tim tells the story of us having to walk to our accommodation on arrival as there wasn't room in the Land Rover for us, it being full of food and stuff. So we walked and talked and got to know each other a little on the walk of about four miles. Tim's version is that neither he nor I were keen to let the other gain a lead in the walk and that in 'keeping up' with each other we both increased our pace until at the end of the 'walk' we were both practically jogging into base. I can't remember that – but it sounds just the type of thing he (and I) would do!

Our base for the summer looked north across Mull to Skye and the Cuillins in the distance. To the west was Canna. That summer was very hard work but great fun and another step towards a nature conservation career. I also met my future wife Rosemary there but as this is not an autobiography we won't linger on that!

That summer I saw my first otter, was bitten by midges, visited one of the world's largest Manx shearwater colonies 2,000 feet up the mountain of Hallival, learned about the sea eagle reintroduction project, learned to dance an eightsome reel and strip the willow, saw a distant pilot whale, watched an amazing show of northern lights, killed a sickly deer calf with a length of lead piping, heard a quail sing and got to recognise a lot of deer individually by their faces and rumps (a very useful skill to take forward into later life).

But aside from meeting Rosemary, the big impact on my life was that I rediscovered wider nature and was shown the intellectual excitement of doing research. At the dinner table in the simple bothy where we lived and ate the talk was about kin selection, Richard Dawkin's new book (*The Selfish Gene*), status signalling, evolutionary biology and parent-offspring conflict. All of this sounded very new and very exciting.

I had thought after my first year at Cambridge that I would do molecular biology and biochemistry in my second year, but a summer on Rhum changed all that. I persuaded Tim Clutton-Brock that rather than pay my train fare home he should give me the money for it (actually he negotiated me down to only part of the money) and I would hitch all the way back to Bristol. Hitch-hiking was a common and effective mode of undergraduate travel in those days and I was quite accustomed to and adept at it – I had pulled the same trick at Easter by hitching to a geology field trip on Arran and using the saved rail fare to fund my Easter vacation.

Mallaig to Bristol was a bit of a stretch but I did it and trotted off to George's bookshop at the top of Park Street to buy my own copies of *The Selfish Gene* and E.O. Wilson's *Sociobiology*. And I changed my study options to ones which allowed me to pursue an interest in whole animals not bits of animals.

I spent the next Easter vacation on Rhum again but the summer of 1978 saw me leaving Cambridge as soon as my Part1B exams were done to head off to southern Norway as a member of the 1978 Cambridge Norwegian Expedition. There were eight of us in all led by John Innes who was a geographer from St Catherine's. John was a member of the (somewhat infamous) Tay Ringing Group and we were all going off to ring and colour-ring purple sandpipers and dotterel. The only non-Cambridge member of the expedition was Keith Brockie, then a young artist and now one of renown.

While our peers were dancing at May Balls we were camping in sight of the Hardangerjokulen glacier on the Hardangervidda plateau in southern Norway. Each day we would set off in pairs to look for wader nests and record what we found, ring chicks and catch adults in nest traps. The plateau was an amazing place of lakes and snow with deep valleys with marshes in their floors. There were plenty of new wildlife wonders to see.

It was a vole year, which meant that we saw lots of lemmings scurrying through the vegetation. And there were plenty of rough-legged buzzards and long-tailed skuas nesting and taking advantage of all that cute rodent food. We saw no snowy owls, wolverines or Arctic foxes though. But there were reindeer in big herds and 3,000 came very close to our camp one day.

The passerines were slightly dull except I did see bluethroats and Lapland buntings for the first time. But the waders were amazing with wood sandpipers, red-necked phalaropes and great snipe alongside more familiar redshank, dunlin and common sandpipers. I would vote for wood sandpiper chicks being the most gorgeous I have seen and Geoff Sherwood and I sat on the edge of a great snipe lek for several nights recording the behaviour of the displaying males.

We ate badly but had a great time and we ringed quite a few purple sandpipers, although not as many as we had hoped. Purple sandpipers, or 'purps', like many waders have evolved a distraction display to lead predators away from their nests if they are disturbed. Whereas dotterels and other plovers will feign injury and trail a wing to lead a fox a few metres away from their nest before flying off perfectly ably the purp 'uses' another ruse: it scurries away with its head held down and its tail low looking like a scurrying rodent. That year the plateau was alive with rodents but each time we saw the 'purp rodent run' it fooled us for a split second, just as it's difficult not to be distracted by a broken-wing display. I'm not surprised it works on foxes.

On that icy high plateau, with the glacier in the distance and shore larks singing around us, I occasionally thought of that first purple sandpiper sitting on the rocks at Sand Point and realised that I was completely hooked on birds and nature. I had moved from being a keen collector of new species to another stage: I wanted to see more of the natural world but I also wanted to understand it and study it.

The observations we had made on great snipe leks were written into a scientific paper and I presented the results, such as they were, at a student conference at the Edward Grey Institute in Oxford the following New Year. I was one of the few undergraduates giving presentations as most were PhD students and I must have made an impression as John Krebs came up to me afterwards and said nice things about the talk. I wasn't quite as impressed by this as I might have been – but I was quite impressed. John was an up-and-coming – actually he had probably 'arrived' by then even though he was to keep rising for many years yet – biologist who, with Nick Davies, had produced a textbook on behavioural ecology, the new trendy subject which looked at how well-evolved animal behaviour was.

I must have said some of the right things because next autumn, after I had failed to get some PhD studentships for which I had applied and turned down a couple of others that didn't appeal, John got in touch with me and offered me a job as his research assistant for a year to finish off the grant left over from when his post-doctoral assistant Richard Cowie had gone to a lectureship in Cardiff.

This was a big opportunity and I grabbed it with both hands. But first I went off to Portugal for a month to study pied flycatchers in a cork oak wood with Colin Bibby and Rhys Green – both of whom worked for the RSPB. I can't remember how this happened, it must have been a conversation at a Cambridge Bird Club meeting I guess, but I do remember being met at Lisbon airport with John Innes (he had flown, I had trained it) and being driven around to the other side of the Tagus Estuary where we slept in a ruined farmhouse next to a beautiful cork oak wood. The woods were full of migrant birds fuelling up before heading to north Africa or perhaps straight to the other side of the Sahara. Our job was to catch and recapture pied flycatchers to look at how quickly they put on weight. For me, it was just fun! I hadn't been to Portugal before and I hadn't seen a cork oak wood either. Our evenings were spent drinking copious amounts of cheap red wine from a plastic carafe which held many litres and cost many pence. As we sat, ate and drank we looked west over the Tagus towards the lights of Lisbon and the setting sun. Far out in the estuary there was a thin pink line of flamingos and in the fields nearer to us black-winged kites would hunt. Just below us in a reed bed was a roost of azure-winged magpies and we would watch them coming in to the roost through the last hours of daylight.

As we sat Rhys and Colin, both to be my colleagues at the RSPB in later years, would pick apart the careers of famous ecologists as ruthlessly and cleanly as a great grey shrike took to decapitating pied flycatchers in our traps. We were using spring traps, baited with large, winged ants, to catch the birds. The ant would be attached to the trap with twisted copper wire and when the bird pulled on the ant the trap was released and the bird caught in a net. But the shrike got the habit of visiting the traps and eating the caught birds until we caught it, painted it pink with the dyes we were using and took it down the road before releasing it.

Repeated weighing of an individual bird at the same time of day was the key to getting the knowledge we sought, and so it was good to catch a bird once but much better to catch it twice, three times, or more over its stay. There was one pied flycatcher, we dyed them on their throats, wing bars and under tail coverts, known as yellow-blank-yellow who stayed in the area over a long period but which we had only caught the once. We wanted its weight again and while the others talked about options I took a handful of traps and said in a Captain Oates-like way that I was going out for a walk – and I wasn't coming back until I had re-caught yellow-blank-yellow. Luckily a surfeit of traps with their winged ants glinting in the sun proved so tempting for the little flycatcher that we got its weight again in the end.

Oxford

Arriving in Oxford after a month in Portugal I reported to John Krebs and he gave me some jobs to do. First though we went up on to the roof of the stark concrete building on South Parks Road which houses both Zoology and Experimental Psychology Departments to visit the aviaries which held some captive great tits. These birds participated in foraging experiments where scientists tried to figure out whether the birds made optimal decisions when faced with simple tasks such as selecting food items from conveyer belts. John led the way and told me to be careful to shut the door of the cage. As we entered the outdoor aviary the great tit inside flew past us both and out of the door which had swung open behind me. I wanted the earth to swallow me up – not only did I not understand the intricacies of optimal foraging theory, I couldn't even shut a door behind me properly. I learned a lesson from John that day – rather than getting cross with me he smiled and said 'That's a pity – that was Ric, he's one of our best great tits', before going on to tell me about the work of Ric Charnov, the American mathematical biologist after whom the departed great tit had been named.

Oxford was an exciting place to be. In the Animal Behaviour Research Group there were figures of renown such as Richard Dawkins, whose energies at that time were still being channelled into understanding the evolution of behaviour. The Edward Grey Institute for Field Ornithology was led by Chris Perrins who had worked with the Institute's founding Director, David Lack, and had subsequently taken forward the study of the great tit population in Wytham Woods. John Krebs had one of the largest research groups with some looking at optimal foraging of great tits in the laboratory and others at the details of great tit song in the University Parks and Wytham Woods. It was a time of ideas and Oxford then attracted the cleverest thinkers and most-skilled doers, bright theoreticians and gifted naturalists. For a young man interested in nature, the emphasis on evolutionary biology added an intellectual interest to an aesthetic interest in birds. Birds weren't just amazing to see and to watch but one's amazement could be sharpened by intellectual curiosity about why they did things down to very detailed aspects of behaviour. We, and it did feel like a great team effort, were working out how natural selection, the great shaper of the life-forms around us, had determined the intricate details of the things that you could see animals doing in their daily lives.

My first main job was to do an experiment looking at whether great tits were more successful at catching live invertebrate prey items at higher or lower temperatures. You can imagine that invertebrates, such as bugs or spiders or beetles, being 'cold-blooded', become more active as temperatures rise. Does that make the task of an insectivorous gleaner like the great tit more or less difficult? Does a bird like the great tit need its prey to be moving so that they are easier to spot, or does the movement and greater agility of the prey make its life more difficult?

Our interest in this question was not just because it would be interesting, but as part of a study on why there is a dawn chorus. Why do lots of birds, of lots of species, in lots of parts of the world, sing in the morning? Is it a mistake or a quirk – surely not as singing takes time, takes energy and feasibly increases their vulnerability to predators? In shorthand, surely a dawn chorus must have evolved, so what are the reasons why singing then is a 'good idea' that has evolved through natural selection?

Various studies showed that the acoustic properties of conditions at dawn (the temperature, the usually greater stillness of the air, etc.) suited the transmission of sound signals at the frequency of bird song – so maybe part of the explanation was that it is a 'good' time to sing to get your message across. Another reason might be that you 'need' to sing at dawn to show that you have survived the night and are still defending your territory against those who want to take it over. Another might be that it is a good time to stimulate your partner through the beauty of your song. Or maybe it's a little more prosaic – dawn is a poor time to feed and so if you have to sing, which you do to attract and maybe keep a mate, and to defend your patch, then you might as well get quite a bit of singing in at dawn. Most birds eat invertebrates in spring and so most may be influenced by the temperature dependence of foraging success. And that's where my little study came in.

I collected invertebrate prey (actually spiders from Oxford's multi-storey car parks) and released them into a cage about a cubic metre in size, in the basement of the Zoology Department, at different temperatures and then a few minutes later released a great tit into the same cage and watched what happened. In brief, the great tit looked for the spiders and ate them, but ate more of them at warmer temperatures as they were moving around more then.

This is hardly the type of finding that gets talked about for years to come but most of science is like that. It's a small brick in an enormous wall. It forms a tiny part of a big whole with 'Why is life on earth like this?' at the top and 'Do great tits get more or less food at higher temperatures?' at the very bottom. The small detailed question is interesting in itself (well, I think it is anyway) but it is a little more interesting because it is part of that hierarchy of questions. And that big question at the top, 'Why is life on earth like this?' can't be answered by one crucial observation or experiment. It is answered by a number of world views which include religious views but also include the theory of evolution by natural selection, Darwin's greatest legacy to us.

Of course, what a few great tits do in a cage in the basement of an Oxford University department is a bit of a step from what is happening in the real world (there is much that happens in Oxford that is a bit of a step from the real world) but only by breaking the big questions down into medium sized, small, and very, very small questions can we get finer and finer detail, whilst always remembering that the fine detail of laboratory work may be slightly or grossly misleading when it comes to the real world. And part of our work was investigating the details of the real world.

If you are lucky enough to have a marsh tit visiting your bird table, watch it and you'll see that it takes the food away from the table and keeps returning to take away more sunflower seeds or other small morsels. Marsh tits are a lot smaller than great tits and always seem like the nice guys of the tit world, but a bit too nice for their own good. Perhaps they take the food away to eat it unmolested. And that is what they do some of the time. But if you are lucky enough to be able to watch them very carefully you will find that marsh tits are hiding some of the food they remove for later. They are food-hoarders.

We wondered how long marsh tits left their hoarded food and how well they remembered where they had hidden it? We wondered whether there were memory tricks to retrieving the food that had been hidden. We wondered all sorts of things.

To investigate marsh tit food hoarding in the real world we developed a clever technique. We set up feeding stations in Wytham Woods in marsh tit territories and got the birds accustomed to feeding at them. We then set up a marsh tit food dispenser which consisted, if you can imagine it, of a rotating vertical disc within a box. The outside of the disc had lots of chambers into each of which we could put a numbered sunflower seed. The disc could be operated from a distance (at the end of a bit of wire) so that we could sit watching the feeder and press the button to release a sunflower seed when a marsh tit arrived at the feeder. And we would know which sunflower seed it was each time because they were numbered in sequence and we could count! Occasionally a great tit would come in and pinch a seed but because we were watching we could record that. Sometimes the marsh tit would eat the seed immediately but because we were watching we could record that too. The final flourish was provided by painting the seeds in radioactive paint because this meant that we could attempt to find the hoarded seeds by walking around the wood with a Geiger counter after all the seeds had been dispensed.

The radioactivity was very low, and was painted onto the seed shell, which the birds removed before eating the seed anyway, so there was no problem with that and we weren't risking our health either.

So, picture the scene. We – David Sherry, a lovely visiting Canadian academic; Allen Stevens, an undergraduate who became a lifelong friend of mine (and birding and racing companion); and myself – would cycle in to the Zoology Department early on an autumn or winter morning, drive out to Wytham Woods, insert the feeder disc with its previously numbered and painted seeds into the feeder, plug in the power source, retreat into the middle distance and wait for a marsh tit to come and have a look at what we were doing. Then, after a wait, we would dispense sunflower seeds to that marsh tit by pressing a button each time he or she arrived and recorded whether the seed was eaten or taken away. When we ran out of radioactively painted seeds (we could dispense around 50), or the marsh tit lost interest, we would start wandering around the woods in ever-increasing circles with a Geiger counter and sharp eyes trying to find the cached seeds. And we did! That's the amazing thing. This technique enabled us to find lots of seeds stuck into the bark of trees or hidden in patches of moss on the woodland floor – seeds that the brain of a tiny bird had told it to hide in those places. We would then subtly record where each seed was hidden and revisit the stored seeds over the next few days to record their fate. Simple, eh?

We also pretended to be marsh tits. We hid sunflower seeds in the types of places used by real marsh tits and then revisited those sites over the next few days to see what happened to the seeds. And we did this many times using different seed spacings. Our hidden seeds did disappear, presumably taken by small mammals and birds that just came across them. We learned, and the observations were quite convincing, that the wider apart you spaced your seeds the more would survive – maybe because predators looked more closely in the area near any seed that they encountered to see whether there were any more. So a marsh tit has to think about where to hide its food in relation to the food source, and find safe hiding places and space out those places appropriately. There's a lot going on inside that little head.

There is a delight in finding out things about the natural world that no-one knew before. It is great if those things are of great import but it's still really fun even if they are as recherché as the fact that marsh tits take the first seeds that they hoard further away from the food source than the later ones which are hidden closer to the feeder. This result was not what we expected because our models suggested that early seeds should be hidden close at hand so that the marsh tit could rush back quickly to grab more seeds before other marsh tits, other birds or perhaps mammals removed all the food. But there you go – we are investigating nature and trying to understand it and it has no duty to conform to our expectations. It was another brick in the wall, and Allen went on to discover more of the details of this behaviour in his DPhil studies.

PhD

That year at Oxford, 1979–80, was a seminal one. It gave me lots of experience in fieldwork and laboratory studies, but more than that it trained my mind and filled my life with long-lasting friendships. There were research talks, some by we learners and some by the learned. There was a large group of keen, young researchers covering a wide variety of subjects, and there were visiting academics from across the world who would make brief visits to Oxford or spend their sabbaticals there. The mixture was heady – at least it was for a young man hooked on nature and hooked on science. There was probably no better place for me to spend that year of my life and there were, of course, parties and girls and beer and friendships.

But it all had to come to an end as the grant money ran out, and what I really wanted to do, more than ever now, was a PhD study of my own. Working at Oxford had given me access to, and some real input into, lots of studies which had been wonderful for me in getting my name onto a few scientific papers. I learned a lot but I needed to do my own thing and get those three letters, PhD, under my belt.

I was interviewed for my eventual PhD studentship by my eventual PhD supervisor, Paul Racey, in the entrance hall of the Natural History Museum. It was one of those interviews that started with hypothetical questions such as 'how would you do this?' but gradually morphed into a conversation about how I actually was going to do things. I was in! And my PhD study would be *The winter activity of pipistrelle bats.*

Paul Racey was based in the zoology department at Aberdeen University and he eventually became Regius Professor there. His research interests were in the reproductive biology of mammals and he had done ground-breaking research on bats in his own PhD study, years earlier. He was always full of good advice on what to do (and, often more importantly, what not to do), was a great sounding board and was a very supportive and helpful supervisor. But one of his best characteristics was that he just let me get on with things – I hope because he thought I probably knew what I was doing most of the time.

So the nub of my study was to take the fact that people sometimes saw bats flying around on warm winter evenings and turn that into a study. My background in Oxford on optimal foraging and evolutionary biology led me into a mixture of fieldwork and a bit of modelling. For many people the first year of a PhD study is a make or break experience. Bad choices or bad luck can stymie your study, necessitate a rethink and lose you a year of work. I was lucky. After spending October 1980 in Aberdeen I was back in Cambridge in early November walking the area around Coe Fen, just on the southern edge of The Backs, at dusk with a bat detector and my shiny eager face pointing to the sky in search of bats. In November 1980 I spent 11 evenings checking for bats on Coe Fen and found them feeding there on about three quarters of those nights. Into December and January bats were still active, but less so, although they were out and about more often than I had expected. Around 30% of evenings saw bat activity, curious considering everyone knows that bats hibernate through the winter. I was off and had something to study.

If I had chosen a different spot, or been less fortunate, then perhaps that early good start would have evaded me. As it was, I knew that pipistrelle bats, identifiable by their echolocation calls picked up by the bat detector (a small box of electronic tricks which renders ultrasound audible to us) were, indeed, as Paul had surmised, active through the winter period. And it was pretty clear as time went by that they were much more likely to be active on warm evenings than on cold ones. And the sounds picked up on the bat detector allowed me to estimate the feeding rates of these active bats.

Bats catch their small insect prey in the dark, by emitting very loud but very high pitched cries, which bounce off objects around about and can be picked up and heard by the bat (which is why bats have big ears). Bats can tell the difference between a twig and an insect from the sound bouncing back to them and so they can chase a juicy flying beetle or fly. As they home in on their prey the bats emit more and more echolocation 'shouts' (they are very loud – you must be grateful that they are too high pitched for us to hear otherwise many an evening walk would be ruined by the deafening noise of bats) to get more and more information on the location of their prey. The final swoop captures the insect in their tail or wing membranes (like a fielder in baseball), before the meal is transferred to their mouths. And that intense emission of closely synchronised 'yells' creates a feeding buzz which you can hear on the bat detector. A bit of wandering around at dusk listening to bats, counting feeding buzzes and keeping track of how much time you can hear a bat well enough to hear a feeding buzz if there is one, will allow you (or me, actually) to work out the feeding success of bats on any night, and relate that to temperature. It's not quite like great tits and spiders in cages but you get the same result – bats have greater feeding success on warm evenings. Interestingly enough, it's true even on warm January or February evenings when maybe you might have expected there to be few insects around.

Collect a lot of data, do a few sums, and you find that bats come out to feed on winter nights when they can make a profit from feeding and don't often come out when it's too cold for there to be insects aplenty. Isn't evolution wonderful? There was a bit more to my PhD than this but I've given you the punch line. And I spent a lot of time talking to vicars in East Anglian churches, mistnetting bats at roosts, winkling bats out of crevices and being called 'Batman' by churchwardens, vicars, their wives and parishioners.

As I was writing up my PhD study, which I did very quickly I am proud to say, I was reading around the subject and dipped into a book that I had bought from a stall on Cambridge market, *The Diaries of Gilbert White.*

There I was quite surprised to read in the entry for 4 November 1777: 'Further it is worthy of note that when the thermometer rises above 50, the bat wakens, & comes forth to feed of an evening in every winter month' and later, on 3 November 1789: 'Bats appear at all seasons through the autumn and spring months, when the Thermr is at 50 because then the phalaenae & moths are stirring'. I'd been scooped over 200 years earlier by one of the greatest English naturalists. And the vicar of Selborne had it right too – it is 50°F, or 10°C, that is the crucial cut off between profitability and energy expenditure for the pipistrelle. I had the sums to prove it, but the Rev. White had got there first. Oh well!

Camargue days

During my first spell at Oxford, John Krebs and I had started to study bee-eaters in the Camargue. John was involved with the Board of the Research Station at Tour du Valat where Dr Luc Hoffmann, of the Swiss pharmaceutical giant, had first established a ringing station and then a research station on an area of Camargue wetland. I continued visiting the Camargue throughout my PhD – the advantage of working on a winter subject! – and then secured a Natural Environment Research Council (NERC) Fellowship to go back to Oxford in 1984–5 to continue the bee-eater studies.

When you tell people that you spent six summers studying bee-eaters in the south of France they tend to roll their eyes, smile and say something like 'Must have been tough' but, actually, it was some of the hardest fieldwork I ever did. I loved the bee-eaters, the work was very successful and there are many bonuses to be had working in such a wildlife-rich place, but I knew I was working too hard one night when I woke at 2 a.m. with my forehead on the lid of a centrifuge which I had turned on six hours before. Fieldwork can be utterly wonderful if you are discovering new things and getting to understand nature better but combining long hours, extreme weather conditions, masses of unfriendly insect life and paperwork, as well as the everyday tasks of cooking, washing and sleeping can be challenging.

Still, looking back it seems entirely wonderful and I am not seeking any sympathy from you, dear reader, as I know I won't get any.

I had never seen a bee-eater before I arrived in the south of France. They are, as everyone says when they first see them, smaller than you expect but utterly gorgeous. A mixture of green, yellow, red, blue and black plumage with a mellifluous call. Our first interest in bee-eaters was because they feed on large insects – things like bees, dragonflies, beetles and grasshoppers. These are the type of insect that you can see in a bird's beak and identify and therefore there was scope for studying foraging behaviour in the field and testing some theoretical models.

In my first few weeks in the Camargue I spent a lot of time watching bee-eaters, getting to know them, finding their colonies, learning their ways and trying to figure out what made them tick. I had arrived at about the time when the bee-eaters were also arriving – they were returning to their colonies in river banks, or occasionally at ground level, where they would renovate their nest tunnels or dig out new ones. The nest chamber would be about a metre down a narrow tunnel and that is where the eggs, up to six or a few more, would be laid. But before the eggs were laid, and while the clutch was being completed, pairs of bee-eaters would sit on nearby telegraph wires and the male would fly out and catch insects – the insects which we were keen to be able to identify. On returning to its wire the male would dash a dragonfly to death with a small number of bashes against the wire and either gobble it down itself or lean forward with its neck stretched low and offer this food to its mate. She would accept the offering on 99 occasions out of 100 and the male would look for the next passing insect to pursue, capture (sometimes with an audible 'snap!' of its bill) and then repeat the whole performance. If a bee was captured, it was its rear end that would get most attention as bee-eaters seem to be rather adept at de-stinging bees with a quick smearing wipe of the bee's abdomen on their perch.

At any time there would be pairs, and very obviously pairs, of bee-eaters strung along the wires over insect-rich meadows with the males dashing out and the females sitting still (and making eggs). The females, I noticed, were just a little greener on the wing than the males and so, often, and certainly with the extremes of both sexes, it was possible to sex a lone bee-eater by its plumage colour.

Because I was watching the birds I started recording their food capture rates, and because I was doing that I thought I ought to record the size of the prey, and because I was doing that I also recorded whether males preferentially offered big prey (dragonflies) to their females or small prey (bees). And I also started recording the handling times of the different prey sizes and how long it took to give the prey, of different sizes, to the female. Occasionally females would hunt insects themselves but they were far less successful than the males – maybe because they were heavy with eggs and insect-catching, particularly the way bee-eaters go about it, is a very athletic exercise.

Male bee-eaters tended to give more of the large prey to their females than the small prey – why might that be? There might be all sorts of reasons but we tested a few alternative foraging models with our data, thanks to the mathematical wizardry of Alasdair Houston who had never, as far as I know, seen a bee-eater in his life. And bee-eaters seem to be getting it right – no surprise there again, evolution is wonderful! They behave as though the pairs are satisfying the energetic needs of the male through the day but maximising the energy provision to the females (who are turning those large dragonflies into eggs for both of them). If males ate more of the large prey themselves this would slow them down too much because 'giving' is quicker than 'eating' when it comes to the large prey whereas 'eating' is marginally quicker than 'giving' when it comes to the small prey. So handing over those large juicy dragonflies represents evolved, informed self-interest for male bee-eaters rather than politeness, gallantry or chivalry. That's evolution for you. And all learned from taking an informed interest in bee-eaters sitting on a wire.

Later in the season a bee-eater colony is frantic. Males and females feed their young and bring insects, one at a time, to the nest. In and out they go and the whole colony is buzzing with activity. At times the parents have to forage far from the colony to find enough food. What sort of prey should they bring back and which sort will they eat themselves? This is a problem of central-place foraging where items are being brought back to a central place – in this case a nest. If you are very close to your nest then it's worth popping back with prey of whatever size, because it's a short journey. But if you are far away, and it takes quite a while to fly home then maybe you should eat the small prey yourself and only travel back with the large prey that you catch. That's what you'd expect from the mathematical models and that, in essence, is what bee-eaters do (not perfectly, but pretty much).

The focus of our study turned more towards the social behaviour of bee-eaters and less about their foraging behaviour. Various interesting things came to light. Up to half of the nests at bee-eater colonies are attended by more than two birds – the additional bird to the putative parents is usually a male. The extra birds act as helpers – bringing food to the nest to feed the chicks in similar ways and sometimes to the same extents as the parents. What was going on here, then?

We discovered that nest predation rates at bee-eater colonies were quite high – from mustelids and snakes (you don't want to put your arm down a bee-eater hole without having a good look for Montpelier snakes first) and that if this happened early in the season the birds would attempt to re-nest. If eggs were lost later on they would give up and sometimes start looking around the colony and visiting other nests, sitting at the entrances and, apparently, looking and listening. After a while a bird would find the 'right' nest and begin, slowly at first but then with greater activity, to 'help' at that nest. We discovered that the helping birds were always closely related to one of the adults at the 'helped' nest (and therefore to the young whom they were feeding). So why were the helpers almost always males? That was because, like many birds, in bee-eaters the females disperse further away from 'home', in this case the natal colony, before they nest and so you are, as a male, more likely to have relatives around you where you nest than if you are female. Added to which bee-eaters don't live long so that it tends to be your dad, your immediate brother or your last year's son that you end up helping – with a very few mums and sisters thrown in too.

So that all makes sense, although we did also notice that there was a fair amount of extra-pair copulation going on, out on the wires, in the spring. Sometimes a male whose female was incubating would give some dragonflies to another male's partner and sometimes that would lead to mating. We didn't know whether that led to successful fertilisation of eggs but if so then it muddied the kinship waters of the colony. As did the occasional case of bigamy – we only saw one – where over a period of days a male chased off the male from the adjacent nest and courted, fed and copulated with a second female – rather blatantly, I thought, in front of his first wife. Once the nests hatched, the bigamous male paid no attention to the second nest but the eventual outcome was that both nests failed, as did many others, in a period of poor weather.

Studying bee-eaters was hard work but very rewarding. Again, it was the mixture of insight into how nature works alongside the intellectual excitement of trying to understand what is going on as well. And bee-eaters are the most beautiful of birds with their busy colonies, intricate social behaviour, and exciting food gathering behaviour.

But then again the people were interesting too. Much of the bee-eater work was done with Kate Lessells, one of the brightest ecologists I've ever met, and a very good friend. John Krebs and I adopted an optimal rule of thumb with Kate. Rules of thumb are what you would expect well-adapted organisms to have. Rules like – 'Feed the big insects to your female' or 'When feeding distant from the colony eat the small insects and bring the big ones back to the nest'. Our rule of thumb was 'Kate's always right' which served us well when it came to household chores, how to do the shopping in the *hypermarche* and the details of evolutionary biology.

Various colleagues, undergraduates and passers-by, helped with aspects of the work – people like Bob Hegner, Alex Kacelnik, Marion East and others – and friends and acquaintances would pop in as the Camargue was a nice place to visit. It was fun to go to the beach for a swim, visit the market in Arles on a Saturday morning, eat *tellines* followed by steak at the local restaurant or do a spot of bird watching in the Camargue, the neighbouring hills of Les Alpilles or the plain of La Crau. It wasn't all work, but we worked hard and played hard and the work came first. Woe betide the person who missed their early morning stint through oversleeping after a night on the *pastis* and *vin rouge*.

The Camargue was a source of many wonderful new sights and sounds. There were the white horses and black bulls but also the wild boar with their humbug-striped piglets. Bonelli's eagles and lesser kestrels still inhabited the Alpilles at that time as did blue rock thrushes. Pin-tailed sandgrouse mixed with stone curlews in the stony desert of La Crau, and the saltpans were where the flamingos nested, and where rare waders such as broad-billed sandpipers also turned up. In the Alpilles at dusk we would see eagle owls and as night fell the tree frogs and nightingales would fill the air with songs.

How I got into the RSPB

Joining the RSPB was a lucky accident. I didn't apply for a job but was offered one anyway. And I'd just told the person who offered me the job that he had fouled up in a scientific paper he had written – not a conventional route into a 25-year love story.

At the time I was based in Oxford at the Edward Grey Institute but the money for my two-year fellowship had ended. As a stop-gap I was doing some work for the Forestry Commission (FC) helping Steve Petty produce a review of birds and forests which was eventually published as a Forestry Commission Occasional Paper.

I had a three-month contract with FC which involved a huge box of papers arriving at my desk and my being told to make sense of them and write them up. One of the papers was a draft on the impact on bird populations of small patches of broadleaved trees in predominantly conifer woods – written by the RSPB's Head of Research, Dr Colin Bibby whose Dartford warbler's colour ring combination I had read many years earlier and with whom I had caught pied flycatchers in Portugal.

I read Colin's paper and found a logical error, quite a big one, in the analysis. After mulling it over for a day or so to make sure that I hadn't got it wrong I phoned Colin up and sketched out what I thought he'd done wrong. 'You'd better come over and talk to me' was Colin's response and so a few days later, I think in February, I was sitting in his office wrestling with the logic of his paper.

When I say wrestled, I don't mean that we were arguing about it – we were both trying to understand it and get it right. Non-scientists might find it odd that Colin was very pleased that I was pointing out his error – well, at least quite pleased! Certainly pleased rather than annoyed. Much better to get it right first, than to publish it and be criticised at a later date. And I was right, and Colin was able to correct it before publication and the paper contains a nice acknowledgement of my role.

Then Colin asked what I was doing that summer – it seemed just a polite enquiry – and I told him of my plans to carry on with bee-eaters for another year even though the funding wasn't totally secure. 'How would you like to go to the Flow Country instead?' he asked. A few months before I would have struggled to know that the Flow Country was at the top of Scotland, straddling the counties of Caithness and Sutherland, and was being trashed by large scale afforestation – but luckily I did know that now thanks to the work I was doing for FC (who were involved in the trashing).

After a brief discussion I was offered a two-year contract leading the RSPB's research on forest impacts on moorland birds. The next day I phoned back and accepted. Little did I know that one day I would be doing Colin's job and that I'd see a flock of little auks from the office window. Little did I know that this was the start of a 25-year love affair with the RSPB that would take me to the Azores, Norwich Crown Court, BBC studios, the Houses of Parliament, Madrid, Cliffe, Washington DC, Otmoor and a host of other far-flung places. Little did I know that it was the start of a 25-year period during which, at work, I would see a wild cat, flying fish, most of Europe's roseate terns, a peacock, a dead yellow-billed cuckoo, cranes in cages and peregrines over

Parliament. And little did I know that I would give evidence to Parliamentary Select Committees, move the RSPB into climate change work, do an April Fool broadcast on the BBC Radio 4 *Today* programme, meet a Prime Minister, march on London's streets, persuade the RSPB to buy a farm, be admonished by cabinet ministers, get letters from the Sandringham Estate, write a blog, and play a part in some of the biggest conservation and environmental issues of the day. I knew nothing of what was to come, but if I had, I wouldn't have changed a thing.

As I boarded the sleeper to Scotland, a little after my 28th birthday, a few days after Dawn Run won an emotional Cheltenham Gold Cup, and a few days before West Tip won the Grand National, I was looking forward to the project ahead. If I had looked back, as I can now, I would have seen my transition from the child whose father pointed out green woodpeckers on family walks to the teenager happy with his own company provided that nature was his companion too. There were years of learning birds through the opportunities that came to me at school, becoming adept at bird identification, and finally being bitten by the bug of science and realising that studying nature was a possibility as a career. I had come to see that ecology and ethology could just be intellectual nature-watching, and that this intellectual element increased the wonder and the joy of looking at birds and the natural world. When I saw a pair of birds on a wire I didn't just want to know what they were, I wanted to know what they were doing and why they were doing it.

The next 25 years at the RSPB became devoted to keeping those birds on the wires.

> *You can know the name of a bird in all the languages of the world, but when you're finished, you'll know absolutely nothing whatever about the bird... So let's look at the bird and see what it's doing — that's what counts. I learned very early the difference between knowing the name of something and knowing something.*
> **Richard Feynman**

CHAPTER 2

Flow Country days

The Great Bog of Sutherland had been so kind to me.
Desmond Nethersole-Thompson

On 1 April 1986 I was heading northwards to Inverness to start work as a Research Biologist for the RSPB. I opened the blinds of my sleeper train compartment at 7 a.m., just north of Aviemore, and looked out the window to see an osprey circling over a lake. This seemed like an auspicious start.

I was heading to the very north of the British mainland to the site of one of the most acrimonious land-use battles ever fought over the desirability of large-scale tree-planting on open areas of high nature conservation value. The area, known to conservationists, but to very few others, as the Flow Country, was being transformed by tree planting encouraged by low land prices and valuable tax incentives – and wildlife seemed certain to suffer.

My part in this battle over the bogs was a minor one, collecting data in a study of the impacts of forestry on wildlife, but my work thrust me into a high-profile political and economic debate from which I learned a lot about how the world worked and the means to achieve conservation success.

The Place

But what of this place? What is it like? Simply, there is nowhere else like the Flow Country.

The area straddles the border of Caithness and Sutherland, in the hinterland south and west of Thurso and Wick. John O'Groats is another 30 miles north and east. To get to this part of the world you either take the road north from Inverness, over the Kessock Bridge over the Moray Firth and north up the A9 past Golspie and Helmsdale, or you can get the train which snakes its way up a similar path through Lairg and Golspie. However you get there you'll find that you feel a long way from everywhere else! It's a long way from Inverness, which in turn is a long way from Edinburgh, which feels a long way from London and the south.

Sutherland is a big county which stretches from the Atlantic and those peaks such as Arkle, Foinaven, Suilven and Canisp across the north coast of Scotland to Ben Kilbreck, Ben Loyal and Scotland's most northerly Munro, Ben Hope. East of Ben Loyal and Ben Hope the terrain calms down and the mountains peter out into the peatlands which are bounded by the sea to the north, and the hills of Scaraben and Maiden Pap to the south. You are now in Caithness where the terrain is pretty flat and subsides eastwards into the farmlands which lie between Wick and Thurso.

This is a place to go to get away from almost everything. You can get away from traffic, people, fresh vegetables in the shops, any chance of salads in the restaurants and early morning newspaper deliveries. But what you do find as recompense is a wonderful landscape of wide open spaces which, if you explore it on foot, is peppered with fascinating systems of small pools clustering in the flat peatlands. For this is a huge *Sphagnum* sponge and as you walk over it, as we did with a lighter and quicker step as the summer progressed and we got fitter, you are walking over thousands of years of accumulated peat. The mosses on which you step today will add their bodies to the metres of wet black rotted ancestors below them. Millions of tons of carbon are stored in the earth beneath your feet.

If you look down you see many *Sphagnum* species and the insect-eating sundews as well. As you look down you may hear a distant greenshank shouting out its song in a display flight, or hear the plaintiff song of a golden plover or perhaps the trilling song of a dunlin as it hovers above your head, skylark-like. For this flattish, wildish wetland is, for just a matter of ten weeks or so, the home of high densities of breeding waders. The dunlin that you may have seen wheeling in the sky above a southern estuary alongside thousands of others is now setting up its own territory – like you, it has got away from it all.

Why don't you visit some time? Go and walk out across the moorland and find yourself miles from anywhere sitting by a group of maybe 20 small pools – the *dubh lochans* in Gaelic (small dark lakes). You may see a red-throated diver on one of these small pools but you'll have to go to the larger lochs to find the black-throated divers. The black-throats can find enough food on their breeding lochs but the red-throats commute to the sea to find fish there. Their calls ring clearly through the air, perhaps particularly when it has been freshened by recent rain, as they fly long-necked over the peaty land, and that may be how they got the name rain geese. Common scoters nest on some of these lochs too – I once came across a nest by accident but it will be July before you see groups of ducklings bobbing on the water.

A hen harrier is more likely to drift past than a golden eagle but you could easily see both as well as merlin and peregrine. Short-eared owls are here too. A few small colonies of Arctic skuas can be seen and even the odd bonxie. And who knows what else? I've seen scaup, whooper swan, dotterel, white-tailed eagle and Temminck's stint in the Flows. And others have seen even more exciting birds.

And of course it's not just the birds. There are red deer, the hares are mountain hares and you may see otters too. The causeway across the Kyle of Tongue is a good place to see them but I once was in a car with Roy Dennis when we saw an otter sitting eating a fish almost by the track. It was a fantastic view. Another time, when driving to start the day's survey work, I saw a wild cat walk down a heathery bank and stand on rocks beside a stream. It's a good place to see adders and lizards as well.

There are the lovely Scottish primroses on the coast at Strathy Point and large heath butterflies whizz over the heaths. And, because I bet you are wondering, there aren't too many midges – particularly out on the moors and particularly in those May and June days.

The Problem

Look up from the wet mossy ground and you can see over the peatlands for miles, and whereas until the early 1980s you would hardly see a tree, except for a solitary pine or birch by the railway line, in 1986 the landscape was scarred by huge new forestry plantations. We saw enormous tractors ploughing up the peat for new trees to be planted and the wildness of the landscape was marred by industrial forestry. The scale of land use change was enormous and as I surveyed the scene I tried to imagine it before the foresters moved in – I wished I'd known the place earlier.

The landscape and land use had been transformed over a period of a few brief years. The trees being planted now, mostly American Sitka spruce and lodgepole pines, would not be felled for at least 30 years. In Caithness, in the area around Altnabreac with its railway platform where you hold your hand out to signal the train to stop, the trees covered an enormous area, stretching for miles, but they had all been planted in the last three or four years. Why was it, that in this formerly treeless area, large scale afforestation had caught on so quickly?

The answer was that one forestry company had spotted an economic advantage from the existing tax system and was buying up cheap land in the area, encouraging investment from high rate tax payers and planting trees in a blanket over the wader-rich peatlands. Fountain Forestry was the company and they were quite an impressive outfit. Their rich customers included Terry Wogan, Alex 'Hurricane' Higgins and Lady Porter of the Greater London Council. Low land prices and the tax system were the keys to making money here – it was really nothing to do with whether you could grow a good crop of timber. And in fact foresters and conservationists did exclaim that this was only a moderately good place to grow trees. You couldn't get a better example of how an economically rational action could lead to an ecologically disastrous impact.

Our study

Trees and most of the birds of open ground don't mix – or to be fairer they only mix for a short few years while the trees are very small after which the bird community gradually changes. You lose the dunlin, greenshank and golden plover, lose the red grouse and lose the hen harriers and merlins. After a while, as the trees grow, you may gain chaffinches and tits, and you may even find some redwings nesting in mature conifer plantations, but the exchange seems to short-change nature at the expense of commercial interest – or, more precisely, tax-savings.

Few really quibbled about the likely loss of open-ground species but there was also concern over whether the mere presence of trees might dry out the surrounding ground, attract predators and change the ecology of the area sufficiently to add to the losses of open-ground species. Some really good wetlands and lochs were surrounded by plantations crowding in on them and it didn't take much imagination to think that this might affect the scoters nesting on the lochs or the waders nesting on the bogs.

And so, in the summer of 1986 I spent most days, with another three field workers, walking 4 km transects parallel to forestry plantations – four of them a day. That's 16 km each day 'on transect' and you could certainly add another 4 km for travel between transects and to and from them at the beginning and end of the day. After a summer of 20+ km days, walking through bogs and over moorlands I was as fit as I'd been for a long time and fitter than I've ever been since.

In subsequent years we tried to find enough wader nests to study predation rates but this was beyond us so we carried out some experiments with dummy nests to examine loss rates at different distances from the forest edge. Our study found some evidence that wader numbers were lower near forest edges, and the more so for older forests, but the evidence was not very convincing then and it's taken a couple of decades more of research for stronger evidence to emerge.

But there was an unforeseen benefit of this work. At the suggestion of Colin Bibby I contacted Dr (now Prof) Roy Haines-Young at Nottingham University – a geographer, no less! Roy, he would happily admit, knew little about birds and would have struggled, like most people, to point to the Flow Country on a map, but he did have expertise in a subject that was a completely closed book to me – satellite imagery.

Now, my natural instinct is to be a bit scathing about geography and that is probably due to all that time colouring in maps with felt tip pens at Bristol Grammar School, but geography had moved on in the 1980s and now you could get a computer to do your colouring in for you. Not only that, but geographers were using satellite imagery to generate maps from space.

The possibility which Colin Bibby had spotted, and which Roy and I followed up, was that we could relate the numbers of birds on the ground to some measure derived from the satellite image. At that time, satellites were passing over the Earth and looking down and measuring the radiation reflected off the Earth in several wavelength bands. That's what your eyes do too, but the satellite 'sees' in different bands of radiation. We wondered whether the numbers of dunlins that we had recorded on our transects might be related to the measurements recorded by the satellite. So we mucked about a bit and then discovered that they were! Rather convincingly actually.

This seemed quite exciting for the following reason – if the satellite data could accurately predict the numbers of waders on the places where we had counted them, then maybe it could predict the numbers on that much greater area on which we had never set foot and hadn't counted them. That would save conservation staff an awful lot of time and produce a map of wader importance for a huge area of land quite quickly.

But one more step was needed – to verify the relationship. After all – the sites we had surveyed were all forest edge sites and might not be entirely representative of the whole of the Flow Country, and maybe we'd just been a bit fluky! Verification was needed. So in 1988 a team of two fieldworkers were sent out to a set of new survey areas which covered the range of satellite data values to see whether dunlin numbers really could be predicted or not. The data would be of some value anyway but if they confirmed the relationship with the satellite imagery then it would be quite a coup.

I was quite hopeful of success but there was one thing that made me have significant doubts about what we would find – Knockfin Heights. The Knockfin Heights is a relatively high (for the low-lying Flow Country) plateau which has lots and lots of pool systems on it. It looks just the type of place where there would be lots of dunlin and that's what the satellite image predicted too. But I remembered Roy Dennis telling me that he had been up to the Heights and there were very few dunlin actually up there. And because three of our survey plots had, by chance, fallen in that area it could be make or break for the value of the relationship between the imagery and wader numbers.

At the end of the season the results showed that the relationship held – very well in fact! The Knockfin Heights plots fitted the expected numbers just as well as everywhere else. We could use the near-infrared band seven measurements from the Landsat satellite in 1978 to predict accurately dunlin numbers in the Flow Country in 1986 and 1988. This allowed us to do what geographers love to do – colour the map in different colours! The colours indicated the good and bad areas for dunlin across around 2,500 km^2 of ground, much of which had never felt the heel of a bird surveyor's boot. Because dunlin numbers are pretty good indicators of the attractiveness of an area for other important Flow Country birds this map was really one of bird conservation interest, so it was quite a step forward.

Our paper was published in the journal *Nature* which was the first RSPB science publication to get into that prestigious journal (actually, although *Nature* is regarded as the 'top' journal one would have to admit there is a lot of luck involved in being published there) and we felt pretty pleased with ourselves. But science moves on and there was a sting in the tail as Roy and I took on a student, the very able Chris Lavers, to take the work further. Chris pushed the study on and showed that although there was nothing wrong with our finding, there was something wrong with our suggested explanation for it. You see, we knew (or at least Roy did) that pool systems, indeed any water, had low reflectance levels on the satellite imagery. And we knew (at least I did) that dunlin liked pool systems. So Roy Haines-Young and I put two and two together and made five by assuming that the satellite reflectance levels were reduced by the abundance of pools on the ground. But Chris showed that it was more something to do with the land in between the pools than the coverage of pools per se – possibly the sogginess of the ground was a good indicator of dunlin numbers but don't quote me on that – or someone will show I'm wrong again.

The People

Soon after I got off the train on that first April day I was in the office of Roy Dennis, the RSPB Highland Officer and Regional Manager whose simple and untidy office was then on the Black Isle north of Inverness. Roy was, and still is, an impressive guy. His knowledge of nature was immense and gained first-hand rather than from books. He seemed to know everyone and every rare bird in his domain. But he wasn't a local, and I used to smile when I heard him say 'we crofters' in his thick Hampshire accent on the basis of his marriage to a local girl.

Roy is a bit like Marmite – you either love him or hate him and although I never hated him I think I have grown to admire him more and more as time has gone on. He is a visionary and, if anything, was at least a decade ahead of his time in the RSPB. Some of what Roy was talking about in the mid-1980s with regard to future land use policy seems more sensible now than it did to many of us then – and that has to be our fault not his.

I also received a very useful piece of advice from Roy about dealing with memos. In those days, they seem quite distant, there was no computer on every desk and mobile phones were not ubiquitous. So messages were relayed on coloured bits of paper and copied to the recipients who needed to receive them. I seem to remember the RSPB memos being a nice light blue colour, but I may be mistaken. Roy's advice was to throw away the first memo you got on any subject *without* reading it. If it's important enough 'they' will send you another memo which he suggested one should throw away *after* reading it. Only when a third memo arrived should one take any notice of it and take it seriously. I have a feeling that it sometimes took a fourth or fifth memo to pin Roy down.

Although I was employed by the RSPB, our work was funded by the

Forestry Commission and so I spent quite a lot of time liaising with them and in particular Arkle Fraser who was based at their small office in Dornoch. Arkle, named after the mountain not the eponymous racehorse, was a Sutherland man and anyone whom Roy did not know Arkle would. We needed forestry maps, keys and advice from Arkle and he gave them all with good humour. He also always provided a delicious cup of tea, with biscuits, in a china cup and saucer so our visits tended to coincide with tea times, and were bracketed with quick looks at the nearby Loch Fleet to pick out the king eider from its commoner relatives.

Further up the road in Wick were the offices of Fountain Forestry who were the main forestry company working in the Flow Country. Their man on the ground was George McRobbie and I used to spend some time in his offices, again picking up keys, forestry maps, contacts and getting advice. Fountain Forestry were seen as being the villain of the piece by the RSPB, since it was mostly their efforts that were opening up the Flow Country and ploughing the peatlands before tree planting began. Not for the last time I was meeting someone a bit suspicious of the organisation for which I worked. And yet, I had to get on with him in a professional way. I don't remember this being a problem with George at all – although I don't have the same strong memory of welcoming cups of tea that I recall from Arkle. Meeting him was one of the first times, and not the last, that as I walked into a room I was carrying the baggage of being 'Mark Avery from the RSPB' rather than just 'Mark Avery', a burden more considerable at some times than others.

To do our survey work we needed access to land adjacent to forestry plantations. This land was sometimes, but not always, owned by the FC or Fountain Forestry, but where it was not then we needed to write to landowners for permission. And it was already early April and surveys needed to start in early May at the latest.

Most landowners were very cooperative. Admittedly there wasn't a lot of damage we could do to areas of rather bleak heathery moor, and if you were just a tourist you would feel free to walk anywhere (even in those days). But as the RSPB, we needed to do things properly.

We received a very witty reply from one landowner, who shall remain nameless for reasons that will become apparent, who in response to our request to make three visits to a part of his large shooting estate replied in the affirmative with just these provisos – that we did not start work until May, that we give him two weeks' notice of our planned visits and that the wind must be from the east. Roy suggested that the first proviso was to give the estate time to bump off any raptors they were targeting and that the combination of the other two conditions was to make the whole thing impossible. But even when we did get permission it didn't mean that everything was plain sailing. Unbeknownst to me, one afternoon as I was wandering across the Flow Country counting birds the owner of a large estate was slagging off the RSPB in the House of Lords for being on his land some days earlier without permission. In fact we had permission from his Factor but his Lordship took exception to our presence.

Solutions

Back in 1986 there was still a fight on about forestry in the Flow Country. Our work was helping to demonstrate the importance of the area for wildlife and the Nature Conservancy Council, as it was then, was also producing dull and worthy reports documenting the area's importance.

The RSPB attacked forestry in this place on all grounds possible – the reduction of wildlife value, the poor value for public money of the tax incentive, the futility of trying to grow trees in such a poor environment, the exaggerated case of the foresters regarding local job creation, the resulting acidification of salmon and trout rivers and a whole bunch of other angles. The public, most of whom, of course, had never been to the Flow Country, never heard of it, and weren't really that bothered, were persuaded that their wildlife should not be destroyed by planting trees that wouldn't grow well just so that rich people could get rich out of tax breaks.

In 1989 the then Chancellor of the Exchequer, Nigel Lawson, announced a change to the forestry tax system that removed the perverse incentive that led to ecological destruction of this sensitive area. But the trees were still standing. Having been planted they grew, albeit many of them not that well, and continued to dry out the peat beneath them, and their impacts reached out into the unplanted bogs around about.

So what to do? When I started walking the Flow Country the RSPB owned no land there but now this area holds the largest single chunk of the RSPB's nature reserve holdings. Some parts are on areas which were never planted with trees but others include areas which were. A massive habitat restoration project is underway here with peatland drains being blocked up so that they stop bleeding the life out of the wetland and trees being removed so as to restore the open spaces and treeless vistas. Millions of trees have been removed in what is really a landscape-scale habitat restoration project. And that map of dunlin densities, and overall bird conservation value, has steered the strategy behind land acquisition to this day.

'Landscape-scale' is one of the new trendy terms in nature conservation. Everyone says it but few actually do it – and few do it on the scale of the RSPB in this distant land of the Flow Country. This nature reserve measures more than 16,000 ha – that's more than 160 square kilometres, or about 55 square miles. That's big! And the RSPB land is spread across a wide area that straddles the Caithness and Sutherland border. It's a remarkable change achieved within the period that I worked for the RSPB. From fighting an advocacy battle (and winning!) to becoming a major landowner in the area and doing a huge amount of practical, valuable conservation work in ways about which other conservation organisations can only dream. If only this project were in the Home Counties so that it were a bit more obvious!

Lessons

I learned a lot from that first RSPB job back in the mid-1980s. I learned from Roy Dennis what you should do with the first few memos you get and I learned about the natural history of a wonderful place from being there, walking across it, seeing the birds and the plants and the mammals and insects. I learned a little of its history and of the Highland Clearances from reading about the area's history. And I learned a bit about interacting with landowners, foresters and nature conservationists too.

As a scientist I learned a bit about working with geographers and how others' expertise can add to your own with surprising effect. I learned to programme a computer to do Principal Component Analysis. But I learned about nature conservation too. Four of those lessons are worth recording here as they have general relevance.

First, information is important. The Flow Country wasn't fully protected because the data weren't there to show how important it was. There are now probably rather few such places on UK soil because the survey work has been done, but back in the 1980s the Flow Country was a chink in the armour of nature conservation. These days the chinks are often offshore – out at sea there may be many Flow Countries of which we know too little and are even now being exploited for oil or gas or wind power.

Second, and related, is the point that information must be translated into site protection that really works. This is now in place in Caithness and Sutherland and the Flows are protected through both domestic and European Union legislation. This will help fight future battles – and there will be some. Since the era of tree-planting ended the Flows' wildlife riches have been threatened by the possibility of dumping radio-active waste and by large-scale windfarm developments. Getting the site protection in place has helped reduce the danger and impact of such proposals and developments, but who knows what may come next?

Third, the power of good and bad policies can be huge and so government policies have to be a focus for wildlife conservationists. No-one intended a single tuft of down on a greenshank chick's body to be harmed by the tax regime governing forestry. Few could have envisaged the impact on this corner of the far north of Scotland through the details of tax relief for tree planting – and yet it was forestry and tax policies which led to the destruction of wildlife sites here. It was changing those policies that saved the Flow Country from becoming a wildlife-poor, forestry-poor tree factory. All sorts of government action have profound impacts on wildlife and nature needs conservationists to be wary and aware of them.

Fourth, putting things right takes a long time and is best done if you have a real stake in the land in question. Whereas stopping afforestation in the Flow Country could be done by an organisation based far from there, getting the land management right is a long term commitment and that is best achieved through land ownership, being a local player and working with local people. Piecing together a large nature reserve up in the Flow Country will help to protect, restore and revitalise this wonderful wildlife area.

And finally...

Since those days of tramping the bogs, day after day, in 1986, I've been back to the Flows a few times on RSPB business. May and June are the best months, to my mind, to visit and you'll see most of the birds best at that time. I never get blasé about the place – it's a wonderful area.

Go and visit! If you do, then get out of your car and walk. Explore – there are not many naturalists who come here so you might break new ground by finding something unusual. Admire the peat-forming mosses beneath your feet and marvel at the beauty of the landscape that stretches away from you. Love the richness of the Flow Country's wildlife and thank the RSPB for what it has done and is doing to protect this very special place.

> *In wilderness I sense the miracle of life, and behind it our scientific accomplishments fade to trivia.*
> **Charles Lindbergh**

Chapter 3

In the pink – roseate terns

I am the very pink of courtesy.
William Shakespeare

I saw my first roseate tern at Dungeness in 1975 when it was spotted and pointed out by Tim Inskipp as it flew over the bubbling water released offshore by the nuclear power station. That distant tern, a bit whiter, a bit longer-tailed, a bit broader- and shorter-winged than the other terns was just a distant view of a new species for me at the time. It made less impact than the spotted crake which we trapped and ringed at the Bird Observatory, and the white-winged black tern and Temminck's stints I saw at nearby Pett Level. Little did I know that roseate terns would take me to three continents 12 years later and that I would get to know this species really well and come to care about it deeply.

In the years after my first view of a roseate tern it became clear, thanks to the work of Alistair Smith, Euan Dunn and others, that it, and most of the terns breeding in Europe, migrated to West Africa where it risked being trapped on its winter quarters. Lots of tern rings were being recovered from countries such as Senegal and Ghana.

During the late 1970s and 1980s it also became clear that roseate terns were declining in numbers in their British and Irish breeding colonies. In the year that I saw my first roseate tern at Dungeness, the British and Irish breeding population was estimated at very nearly 1,000 pairs and yet in 1987, when I had become more fully involved in the roseate tern story and barely 12 years later, the numbers were down to fewer than 450 pairs. The RSPB owned lots of nature reserves dotted around the coast where roseate tern numbers were falling dramatically – places such as Coquet Island, Horse Island, Inchmickery, Green Island and Ynys Feurig - so the decline in the numbers of roseate terns was personal! But these declines weren't restricted to RSPB reserves in the UK – declines were also occurring on sites such as the Farne Islands, managed by the National Trust. And, they weren't restricted to the UK alone – Irish populations were declining too and there were worrying rumours concerning the species in Brittany. As for the Azores – nobody seemed to have a clue as to what was happening there.

In 1987 I put together a research proposal to work on roseate terns and managed to get European funding to bring together all the European states with roseate terns to compare notes and pool expertise. This was a good example of how a species in decline in a political area much bigger than the UK can mean that there is both more money and enthusiasm available to do something about it.

During 1987 I visited just about all of the UK and Irish breeding colonies of roseate terns, talked to most of the people involved in their management and study, went to Brittany and learned about the situation at the few colonies on the north coast, visited the Azores and saw more roseate terns than I had ever seen before, and also spent a few days in the eastern USA talking to those who conserved and studied roseate terns there. The result of all this was a lot of fun for me, a report for the European Community ACE (Action by the Community for the Environment) Programme and, rather more importantly, the start of a more certain future for the conservation of roseate terns throughout their European range.

UK colonies

All roseate tern colonies are different. In Northumberland I visited both the Farne Islands and the RSPB's nature reserve on Coquet Island. This was on 12 June 1987 as I remember sitting up the night before in a bed and breakfast in Seahouses watching the early results come in as Margaret Thatcher achieved her third general election victory. Both the Farnes and Coquet are flat, low-lying islands with little vegetation. There were plenty of Sandwich, Arctic and common terns, but the roseates were much scarcer and at each site they had been much more numerous 20 years earlier. In fact, Coquet Island had been where Euan Dunn had carried out his PhD study of terns in the 1960s when roseates had been common enough that it was possible to study their behaviour in some detail.

Later I visited the Firth of Forth which used to have large roseate tern colonies on islands such as Fidra where they were no longer present at all. Travelling to the RSPB nature reserve of Inchmickery Ian Bainbridge and I found a few pairs of roseate terns nesting on a steep south-facing slope and found some nests and recently hatched chicks. Roseate tern chicks are cuter than those of common terns – with black legs and spiky, punk-like down.

They lay 2–3 eggs to a clutch, which are also longer, thinner and more artistically marked than those of common terns. I am biased, but even if I weren't, I would say that they are an absolutely top bird.

The few pairs that continue to nest on islands in the Firth of Forth are the northernmost nesting roseate terns in the world. I always think of them as I cross this body of water and marvel at the adaptability of species. The roseate tern is the most cosmopolitan of all UK seabirds. First identified in the Firth of Clyde, where they no longer breed, and hanging on in the Forth in small numbers, this bird breeds in the Indian Ocean, the Caribbean and Australasia too. It has a far wider geographic range than its close relatives, the Arctic and common terns, and yet it is far less numerous.

In Wales I visited the Anglesey colonies of Ynys Feurig, Cemlyn and the Skerries. The Ynys Feurig colony had been an important one for a long period of time and was situated on a small island which was accessible from the shore at low tide. This meant that it was vulnerable to accidental disturbance from people and was one of the targets for egg collectors too. As a result the RSPB wardened the site but even when one could keep people off the island it was difficult to persuade the foxes not to cross over and impossible to talk the peregrines out of taking a few terns here and there. In the year I visited a fox did get onto the island and killed at least 57 roosting terns including 12 roseates. In 1980, the wardens estimated that a peregrine took 36 adult roseates at a time when the colony numbered 140 pairs. It's easy to see why there are few such mainland roseate tern colonies and why almost all the colonies are on offshore islands free from ground predators and an annoyingly long commute for avian predators.

Ireland

In contrast to the rather worrying and depressing story from Anglesey, my visit to the Irish colony of Rockabill with the Irish naturalist David Cabot was most uplifting. Rockabill is at the northern end of Dublin Bay and sits a few miles offshore. It's a typical site for roseate terns in some ways. It's an offshore island with large numbers of nesting common terns (almost all the roseate colonies are mixed with larger numbers of common terns). Back in 1987 Rockabill had about 250 pairs of nesting roseate terns and as such it was the largest colony in Europe outside of the Azores (where we didn't really know much about what was happening). When I visited, the numbers had been growing for many years and Rockabill stood out like a beacon of hope for roseate tern conservation in Britain and Ireland. The terns shared the island with a few black guillemots and a team of lighthouse keepers and that was it.

Around the lighthouse there was a walled garden in which grew a veritable forest of tree mallow bushes. And the roseates nested in the garden under the shelter of the tree mallows – almost forest-living terns. The fact that this colony was so successful compared with other north European colonies seemed significant – there must be lessons to be learned here. Do roseate terns need more cover and shelter than other species?

But the critical thing I discovered from Rockabill was that the island's lighthouse was going automatic very soon and that the lighthouse keepers, who had acted as *de facto* wardens for this important site, would no longer be in residence. As a result of my visit the RSPB and the Irish Wildbird Conservancy (now BirdWatch Ireland) got together, with EC (now EU!) support, to ensure the site was wardened and managed into the future.

I remember a conversation with David Cabot as we stood on Rockabill about how many terns the island could hold. My recollection, perhaps inaccurate, was that David thought that there was space for a few more but not many more terns. I was pretty confident that the place could be even more heavily populated.

I was lucky enough to return to Rockabill in July 2006 with some of my RSPB team – it was a fantastic day, with close colleagues and wonderful sunshine. As we travelled across the sea to the island we saw lots of roseates and common terns, and a few Arctics too, fishing around the island. From the boat we watched gannets and Manx shearwaters passing by, as well as storm petrels. As we landed on the island we were greeted by a wardening team from BirdWatch Ireland and by huge numbers of terns! There are now over 600 pairs of roseate terns on Rockabill and large numbers of common terns with a few Arctics as well. What a place! What a success story! What a joy!

Sitting overlooking the colony that day in July you could see and hear and smell that this was a successful seabird colony. And it holds around a half of all the nesting roseate terns in Europe these days and continues to pump out large numbers of young terns to make the journey to Africa and return to nest in their second or third years.

Brittany

The French colonies of roseate terns contained about 100 pairs in a good year and they were situated on the north Brittany coast close to the beaches where half of Paris seemed to assemble in July and August. These colonies faced the usual problems of ground predators – foxes, but also feral ferrets and the occasional peregrine falcon – as well as masses of unintended human disturbance. The staff of the small Société pour L'Etude et la Protection de la Nature en Bretagne did a great job in intervening to prevent too much disturbance of these birds. But I must admit that although we saw the colony from a distance my clearest memory of my visit was eating *crepes* and drinking *cidre* with Breton conservationists who had as much passion for these birds as did their Irish and British counterparts.

The Azores

My taking stock of roseate fortunes also took me to the Azores – Portuguese islands out in the Atlantic, about a third of the way to the New World and volcanic outcrops from the mid-Atlantic ridge where the European and American plates meet. I visited the Azores several times and loved every minute there. For a nature watcher, a good reason to go to the Azores is to see whales, and you can still see the former whaling stations from which Azorean men used to row with harpoons to kill sperm whales up until only a few years before I first visited. And increasingly birders go to the Azores in autumn to see vagrant American land birds and waders. But the oceans surrounding the Azores, crossed by tired parula warblers and bobolinks in autumn, also provide food for nesting seabirds as well as passing whales.

At dusk one can see Cory's shearwaters flying through the street lights of Ponta Delgado, the principal town on the main island of Sao Miguel. One evening, Colin Bibby and I estimated that there were 40,000 of these birds roosting on the sea north of Flores – the westernmost island of the archipelago. On another visit I saw flying fish from a Portuguese naval ship as we travelled to Graciosa. In places the sea was littered with 'plastic bags' that were actually thousands of Portuguese man-of-war jellyfish. We overtook a leatherback turtle and Risso's dolphins swam next to us as we visited small island tern colonies. Here you can find petrels and shearwaters of various species and there is still much to be learned about them.

Island faunas have played a large part in developing our understanding of avian evolution and a trip to the Azores cannot but open up one's eyes. There are fewer species than you would find in a similar mainland area and in similar mainland habitats. Many of the species are familiar – such as the grey wagtail, bullfinch and chaffinch – but there are differences. The bullfinch on the Azores is an endemic species, which was once common enough to annoy fruit growers but is now restricted to the tops of the highest mountains on the Sao Miguel. Similar in looks to a female European bullfinch this is now regarded as a separate species – when did it arrive on the Azores? And how did the bullfinch turn into a new species? The Azores chaffinch is, if anything, more different in looks from its continental relatives than the bullfinch is from its relatives, but is regarded as simply a race rather than a separate species. It actually looks quite blue – as interestingly does the Canaries blue chaffinch stuck on another Atlantic island. And the grey wagtail is the only wagtail on the island and is found everywhere. Although it looks just like 'our' grey wagtail it occupies a much broader range of habitats. It's the grey wagtail you see in gardens, in woods, on the beach, at the top of mountains – everywhere it seems. Is this how grey wagtails 'want' to behave everywhere but are constrained by other species? Is it the lack of many other wagtails, pipits, warblers and flycatchers that allows these grey wagtails to lord it over so many Azores' habitats?

But back to roseate terns – I learned that most of the European population nests in the Azores and although there are a few generally permanent colonies many birds nip around the island archipelago from year to year and are so difficult to pin down. I visited a colony in the east which included a single sooty tern, a colony in the middle archipelago where there were no common terns but little shearwaters and Madeiran storm petrels using the same site, and stood on the westernmost point of the Western Palearctic and saw roseate terns.

I spent a lot of time during that summer and autumn of 1987 looking for and watching roseate terns. I really got to know them well all those years after that first view over the Patch at Dungeness. I became the roseate tern's greatest fan and still am, and although I am biased, roseates, for me, are a lot prettier than common terns in every way. If you see a black-headed gull you might think it is an elegant gull compared with a herring gull, but a common tern will look much more graceful than any gull, and in the same way, and to a similar extent, roseate terns make common terns look like lumbering brutes when you are faced with the impeccable grace and beauty of the roseate tern.

Compared with common terns, with which they are almost always seen, roseates are much whiter and close up they have that pinkish tinge to their breasts which gives them their name. Their wings are shorter and tail streamers much longer.

You can't help but get attached to a species which you study. And you do get to know them very well. I am sure that I would recognise a roseate tern's call even if one flew past out of sight when I was in central London, and one picks up little bits of knowledge such as the fact that roseates seem to me to show their red legs in flight much more often than other terns. I wonder why they do that? They must know they are gorgeous.

A brief visit was enough to confirm that there were plenty of roseate terns on the Azores (and we now know the population is about two thirds of the European total) and to make contacts that allowed future conservation work to be taken forward.

Thinking back to those times, I knew then that it was a great privilege to be able to gad about in pursuit of this fantastic bird and to meet so many lovely people in some amazing places with lots of novel wildlife around me too. But those were the days before the internet (and actually before faxes, believe it or not!) and the personal face-to-face approach was needed. The work I started in 1987 was enough to establish a good case for a more concerted approach to roseate tern conservation across the UK, Ireland, France and Portugal and it kicked off a great deal of good conservation effort by a lot of people. The RSPB appointed Adrian del Nevo to work on this issue and so my focus moved off to other things whilst Adrian did a great job working with the contacts I had made and moving things on very significantly. For example, almost all of the sites where roseate terns nest across Europe are now designated as Special Protection Areas (SPA) for birds under the EU Birds Directive which will protect them (or should!) from the most gross threats.

The USA

In the autumn of 1987 I spent about 10 days in the US talking to people who cared about roseate terns in that part of the world. They had seen declines in numbers too and some of their focus was on where their birds went in winter – somewhere off the coast of Brazil it was thought. I spoke to Fish and Wildlife Service staff and local conservationists from Maryland to Maine.

On Great Gull Island at the mouth of Long Island Sound I heard about tern conservation from Helen Hays who had studied terns there since the 1960s (and still does), saw grey catbirds, ate freshly caught lobsters and was told that I was sitting in the chair that had supported Robert Redford when he visited the island. On Cape Cod I saw a flock of roseate terns at dusk which was about the size of the entire European population, ate clam chowder afterwards and chatted to Ian Nisbet about terns and Kennedys.

There was much useful exchange of information – the Americans were fascinated by what we knew about tern-trapping in Africa, by the fact that our populations were in trouble too and by stories of places where terns were doing well such as Rockabill. And I learned a lot about how similar were the US colonies to our own and that the Americans had done a lot more science on their birds on the breeding grounds. But the most useful thing I got in America was the discovery that roseate terns will nest in nest-boxes!

Americans had noticed that often in mixed tern colonies the commons nested out in the open and the roseates nested under cover. I was told the story about a colony where a winter storm had thrown lots of timber and other rubbish onto a beach and the roseates had all nested in the piles of debris. The Americans had taken this idea and made it real – if roseates wanted more shelter then why not give it to them? And they had provided simple wooden nest boxes in some colonies and the roseates had adopted them with evident alacrity and apparent enthusiasm (birds don't have very expressive faces!). This was the simple idea that I brought back to Europe and it played a major role in management to increase nesting roseate numbers at Rockabill, and also on the RSPB nature reserve of Coquet Island where numbers have increased in recent years. Simple but effective – we need more ideas like these in nature conservation.

Ghana

I also followed the roseate terns to their wintering grounds in west Africa – to Ghana. Building on the excellent work of the Scottish policeman Alistair Smith and the biologist Euan Dunn I made a few visits to the coast of Ghana with Adrian. The RSPB had set up a project with the Ghana Wildlife Society called the 'Save the Seashore Birds Project' which aimed to persuade the coastal people of Ghana to stop trapping terns along their beaches. Ringing recoveries showed that Ghana seemed to be a hotspot for winter tern-killing.

I loved Ghana as a country – and the Ghanaians as people. The country is ramshackle and poor, with many of its people living in shanty towns with little sanitation, but the people I met in Ghana were amazingly welcoming and friendly. At various times I covered the whole of the coast from Togo in the east to Cote d'Ivoire in the west and we would turn up in coastal villages and always be surrounded by hordes of children – smiling and welcoming, not begging or pestering.

But some of those children would occasionally peg a noose on the beach with a dead sardine by it to catch a passing tern that would dip down to pick up the fish and be caught in the noose. Many of the terns trapped were the black terns that fed along the coast in large numbers, and I once saw a tern no sooner caught than its primary feathers were immediately pulled out of each wing by the young child who had secured it. Some of these captured terns were eaten but the major motivation for catching them nowadays seemed to be boredom rather than hunger.

I never saw a roseate tern captured in Ghana but the ringing recoveries showed that they were in fact more susceptible than the other common and much more numerous species such as black, common and Sandwich terns.

In retrospect it looks a little odd that tern trapping was the focus of the work of our BirdLife International partner in Ghana, but this was a consequence of the times and of the early development of the BirdLife partnership in Ghana. There are, of course, much bigger and more important conservation issues for the people of Ghana to face and fix. But as is often the case, you have to start somewhere, and you sometimes have to start with where the funding opportunities are. Tern conservation gave an excellent opportunity to bring local involvement on board and the Wildlife Society of Ghana is certainly now a competent BirdLife partner which does much to stand up for Ghanaian wildlife overall.

The time I spent in Ghana with people like David Daramani (one of the best birders I have ever met) and Yaa Ntiamoa-Baidu were a great education for me. Meeting nature conservationists from a country much less affluent than our own was an eye-opener. We are often told that nature conservation is a luxury that we cannot afford when it stands in the way of economic progress, but to see the passion of African nature conservationists who shared our love of nature despite not sharing our pampered and profligate life-style was a privilege. They cared for the lot of their fellow Ghanaians but they cared for nature too.

And, of course, being in a new country and spending much of each day counting terns and other shorebirds, one sees a lot of other new birds too. Grey parrots (with their red tails) inhabited the coastal forests, jacanas walked across the water lilies and grey plantain eaters flew across the forest roads.

Up and down the coast we travelled visiting saltpans, lagoons and long sandy beaches. This was autumn and the season for sardine fishing. Long wooden boats would head off from the shore to encircle shoals of sparkling sardines that would later be lined up on the beach to be sold. And the terns were here for a similar reason – their arrival coincided with the peak sardine catches and the fishermen knew that if there were lots of terns fishing offshore then there must be fish for them to catch too.

Surveys by Ghanaian staff of the Save the Seashore Birds Project showed that tern numbers peaked in September and October and then dropped off until after Christmas when very few terns were sighted. We still don't know where these terns go then. Do they just change their habits and become a bit more difficult to see, or do they only come ashore long after dark later in the northern winter? And roseate terns come back to their breeding grounds late in the year – in May – so the whereabouts of the European roseate terns is still a mystery for at least a third of the year.

We saw roseate terns at the Densu saltpans outside Accra. Terns would come in to roost there and one could pick out the roseates amongst the other species including Sandwich, common, Caspian, royal and, once, a black noddy. We saw a few UK and Irish colour-ringed birds and the occasional bird with a yellow Azores colour ring on its leg. Once I saw a bird which was almost certainly from Rockabill sitting next to one that had been ringed a few years previously in the Azores. Ghana was indeed the place where Europe's roseate terns met and mingled.

Conclusion

All in all, roseate terns are now doing well in Europe – certainly much better than we had feared back in the mid-1980s. Looking back on the action plan we wrote in 1987 from this distance it looks pretty good. And it might easily be one of the more influential things I did for conservation during my time at the RSPB. It kick-started a lot of site-protection work around Europe – wardening and designation of the most important sites. Most of their breeding colonies are now protected from development, disturbance and, in many cases, from predators.

Overall, their population has risen but is very dependent on continued successful breeding on Rockabill and the Azores. The nest boxes seem to be doing a good job. The worry is, and it always is with species like this, that we have rather too many of our roseate terns in too few baskets so that too high a proportion of the total may suffer from any single local impact. And we know very little about their food supplies. What if the fish around Rockabill all go somewhere else? What happens then? To be honest, we'd just have to keep our fingers crossed and hope that one of their former colonies dotted around the UK would become favoured once again.

Although not many current RSPB staff know much about my role in the roseate tern story it is one of the things of which I feel most proud. We've done well for this bird and I feel glad that I played a part in starting a lot of effective conservation action for it.

The bird itself is still a great favourite of mine. I don't see them very often these days but when I do, my mind goes back to the sound and smell of them nesting in the walled garden on Rockabill, or sitting above a roseate colony in the Azores one night with Adrian del Nevo and a bottle of local brandy and chewing the fat, or the coconut-strewn west African beaches with the terns flying along the shore.

The roseate tern is now somewhat in the pink. It's another successful battle in the war. From this battle I learned a few lessons.

First, international cooperation is essential in conserving migratory species. With such birds their future depends on all the places where they live or visit being managed well. If the Ghanaians are killing our wintering terns then we will lose them, but if we are destroying their breeding colonies then the people of coastal Ghana will see fewer terns too. Nature is a shared inheritance and we need to conserve it together. The EU Nature Directives, which make both Rockabill and sites in the distant Azores SPAs, is an enlightened mechanism for international cooperation in nature conservation – it helped to get French, Irish and Portuguese conservationists all talking the same conservation language.

Second, everywhere you go you will find people who love nature and want to work with you to help it. They may have different religions and ways of living, certainly they may speak different languages from you, but they will be on the same wavelength when it matters. This means that international cooperation is always feasible provided you find the right people to work with and go about your work in a sensitive way. And it is great fun having a network of contacts across the world who care about nature too.

Third, you need to have the information on numbers to understand what is happening. There aren't many roseate terns in Europe and they don't nest in that many colonies but piecing together the population trend depends on getting good and regular information from the most important colonies because otherwise there is always the temptation, particularly with terns which move about a lot, to put down any possible decline to a movement of birds to other places. Such contacts are now much easier to maintain but good information is the basis of any successful conservation action.

Fourth, some species, and the roseate tern is one of them, occupy very few sites which are therefore proportionately very important for that species. An Azorean fisherman is unlikely to tell roseate and common terns apart and even less likely to know that a colony of 100–200 pairs of roseate terns is about 10% of the entire European population, and so the proper identification and protection of such sites is an incredibly important tool in the nature conservation toolbox.

I believe in pink
Audrey Hepburn

CHAPTER 4

Counting, cubes and curves

We have the most crude accounting tools. It's tragic because our accounts and our national arithmetic doesn't tell us the things that we need to know.
Susan George

Any management consultant will tell you that you need to set targets, measure progress and set priorities. While ignoring this blindingly obvious advice can cost you a fortune as a business, it is nowhere near sufficient for complete success.

When I joined the RSPB, the organisation was slowly, and somewhat painfully, learning these lessons. When one is trying to do good things and change the world for the better, it can sometimes be difficult to distinguish the things that should be done from those that should not. Nothing you can do as a nature conservationist is likely to be worthless but much of what you, or someone in your organisation, might want to do may not actually be for the best. So – what is 'the best', and how might we recognise it?

A good basis for judging where to put conservation resources is to know what is happening to species now and in the recent past – which species are increasing and which are decreasing? That's not all you need to know – but it's a good place to start.

Counting

Bird monitoring in the UK is a pretty sophisticated enterprise – it is to some extent the envy of other countries, and to an even greater extent the envy of those working on the vast majority of wildlife not lucky enough to be feathered. Aren't we birdy people lucky? Not really – luck had very little to do with it!

What would a perfect bird monitoring programme look like? Ideally we would have population figures for all bird species every year and in enough detail to look at regional differences. And we are not a million miles away from that perfect position.

Before describing the state of UK bird monitoring, let's just think of the challenges facing the person or organisation who might want to set up a bird monitoring programme to cover all our bird species. Here is a list of difficulties you would have to overcome:

- Some birds occur here in winter, some in summer, and some on migration only
- Some are rare (the occasional pair) and some are very common (millions of pairs)
- Some are nocturnal, but most are diurnal
- There are lots of species – over 300 recorded in the UK each year
- Birds move about
- Some are difficult to see but easy to hear
- Birdsong is often most common ridiculously early in the day
- Weather affects birds' behaviour and therefore their visibility
- Birds come in all sizes and colours
- Some are deliberately cryptic in plumage and behaviour
- Some are highly colonial and are therefore found in huge numbers but in very few places
- Birds occur in all habitats from the centres of towns to the tops of mountains
- Some species are difficult to distinguish from others
- Some have different plumages at different times of year, or look different according to gender or age
- Their behaviours change through the year

Now I'd rather be set the task of counting the UK's birds than the UK's mammals but that task is not simple. Bird-people tend to think that, with its imperfections, bird monitoring is simple because of how well we have mastered the art rather than because it was, of itself, a cinch!

This list of problems to be overcome could pretty quickly lead to the view that no single scheme could cover everything and that you are going to have to rely on mobilising an army of technically proficient volunteers who are skilfully directed by a bunch of clever boffins. And that's the solution we have achieved.

UK birds are primarily monitored by the British Trust for Ornithology (BTO)/Joint Nature Conservancy Council (JNCC)/RSPB Breeding Bird Survey (the BBS), which covers those species widespread in woodlands, farmlands, moorlands and urban areas. Set up in the 1990s it depends on a small army of dedicated volunteers (and I am one of them) who visit randomly selected 1 km square locations twice in the spring and count all the birds seen and heard along two straight line transects across the plot.

I like the fact that while I am recording a blackbird on a dull bit of farmland on the Cambridgeshire/Northamptonshire border, someone else may be recording a blackbird in Dollis Hill while another observer records a blackbird in a birch wood in Sutherland.

Over 2,000 observers take part in the BBS and they deserve a lot of credit for providing fantastic basic information on bird populations. If we could increase the number of active observers, particularly in the less populated parts of the UK, then that would improve the figures a bit – and rope in a few more species whose sample sizes are small at the moment – but given the realities of life, this is a stupendously successful survey which provides most of what we need for many breeding species.

The width of the survey's coverage is important. The temptation is always to focus on species of immediate interest or concern, but the best monitoring scheme will do that reasonably well whilst also covering many other species that aren't today's priorities. If collared doves were to decline in numbers (and they probably are now!) then it's no use wondering how to measure those declines when they've started or are already being reported by members of the public. We need historic information as a baseline and the best monitoring scheme will arm you to deal with tomorrow's problems as well as allow the success of past conservation actions to be charted.

The BBS had a predecessor – the Common Birds Census (CBC) started in the early 1960s after a couple of cold winters had dramatically affected bird populations but at a time when little national monitoring existed. The CBC was great when it was set up, but had a lot of snags. It was more time intensive for the volunteer (who had to make about half a dozen visits rather than a couple) and for the analyst (lots of mapping rather than adding up) – and it only covered woods and farmland. The survey sites were few and largely self-selected by the observer, and so did not represent woods and fields as a whole, and the survey had a south-eastern bias that it seemed would never be reversed.

But the CBC was, for about 25 years, very good. Towards the late 1980s it became more and more obvious that its limitations were important ones. I played a part in getting the BBS off the ground – first in what was seen as a not very helpful way and latterly to more positive effect. The first involved writing a review for the British Ornithologists' Union (BOU) journal, *Ibis*, for a celebratory volume which was produced by the BTO to cover the CBC's results. I looked again at that review recently with a certain amount of dread as I know it stirred things up quite a bit! But I was relieved to find that it seemed to me to fall into the 'hard but fair' category. Certainly the review was novel in putting the CBC's limitations into the 'public' domain.

A few years later, as RSPB Head of Conservation Science, I worked with Colin Galbraith, then of the JNCC, to persuade the BTO to do the work that was needed to move from the CBC to what is now the BBS. And I persuaded the RSPB (largely myself!) that the BBS would be such a great leap forward that we should help pay for it, and promote it. And that's how the BBS became a joint venture – although BTO staff and their Regional Representative network remain the driving force behind the scheme's successful running.

I regard the BBS as a great achievement – largely an achievement of the BTO staff and volunteers. But thinking back to the days of its slightly painful birth the proud parent needed an awful lot of help to give issue, and some of the encouragement came in the form of tough love provided by the RSPB.

Colin Galbraith and I had worked earlier on another birth – of a different sort. We helped persuade the BTO and the Wildfowl and Wetlands Trust (WWT) to merge their existing Birds of Estuaries Enquiry and National Wildfowl Counts (and low-tide wader counts) into one joint scheme called, rather wittily given its focus, WeBS (the Wetland Bird Survey). This is the winter companion of the BBS – counting winter visitors, mostly waders, ducks, geese and swans at estuaries and lakes. It seemed obvious to Colin and me that it was in our organisations' interests, but also in those of the volunteers involved, that two schemes covering similar species at similar times should be merged. It wasn't that simple in practice and the whole negotiation seemed to take forever but we got there in the end and the merged scheme became easier and cheaper to run.

WeBS provides information on the internationally important populations of waterfowl that winter in the UK – all those knot, barnacle geese, wigeon and grey plover. I've never done WeBS counts – too much of a warm-weather birder, me! But the data are used every day by the RSPB in protecting estuaries and other sites from development threats. WeBS is a very strong, and very important pillar in bird conservation.

Seabirds are a bit of a problem – many of them live in remote places where counters rarely go and terns aren't very site-faithful so the value of occasional counts of some colonies is reduced. And then there are the nocturnal species and the fact that many live in very large colonies where counting is a challenge. It's not surprising that we don't have good annual population estimates of Leach's petrels, all things considered.

And so seabirds are assessed by an annual population monitoring scheme backed up by 'complete' counts every 15 years or so. We're heading for the next of these big counts and it's likely that it will reveal big changes in many seabird populations. Given that more than half the world populations of species such as gannet, bonxie and Manx shearwater are found in shores around the UK and Ireland, knowing their fate is important. We'd expect China to keep an eye on the giant panda population, and Australia to assess that of koalas, so we need to know whether gannet numbers are going up or down.

We're now largely into filling in the gaps in our knowledge. The very rare species don't need their own surveys and it is OK to rely on records coming in from the public, birders and conservation organisations to keep tabs on them. The admirable Rare Breeding Birds Panel (RBBP) carry out this task – they'll tell us whether scarlet rosefinches really are colonising the UK (providing we all tell them first through submitting our own secret records). The annual RBBP report, published in British Birds, is a fascinating read every year. It's one for a glass or two of whisky in front of a log fire.

Some species are not very rare, but are too rare to be picked up by the BBS. Many of these species, e.g. hen harriers, dotterel, cirl bunting, etc., are traditionally species of conservation concern. As Head of Conservation Science, I established an agreement with the statutory conservation agencies to agree a timetable of either annual or periodic surveys for these species. Going under the bizarre acronym of SCARABBS (Statutory Conservation Agencies and RSPB Annual Breeding Bird Surveys) it provides a structure for planning resource allocation for bird surveys into the future. Despite unforeseen shocks to the system, such as the Foot and Mouth Disease outbreak in 2001, and occasional lack of money, this programme of survey work forms the backbone for one-off and regular surveys.

This family of surveys (BBS, WeBS, seabird monitoring, RBBP and SCARABBS) forms the framework for dealing with all the snags regarding the difficulties of surveying birds in the UK. It's not bad at all. Looking back, the RSPB's role in all this has grown enormously over the last 25 years. Our intellectual input has been matched with financial contributions to underpin this monitoring and that has got the RSPB logo onto lots of survey work where it never appeared in the past. I'm moderately proud of the role I've played in this but it has been very much a collaborative team effort. Heroes of the growth of bird monitoring over this time include a large number of people in the statutory nature conservation agencies such as David Stroud (JNCC), Andy Brown and Phil Grice (English Nature - EN and Natural England - NE) and Des Thompson (Scottish Natural Heritage - SNH). Stephen Baillie is a superstar at the BTO for his role in designing and running the BBS. Mark Tasker (JNCC) has been a moving force behind much of the seabird monitoring for decades. At the RSPB, David Gibbons and Richard Gregory have, through the years, played major parts in establishing and running SCARABBS and managing the RSPB's considerable financial input into BBS, WeBS, seabirds, RBBP and SCARABBS.

Bird population monitoring is so fundamental to identifying priorities in nature conservation that helping BirdLife International partners to set up their own bird monitoring schemes has become a standard part of the RSPB's international work. Across Europe and beyond, RSPB staff have helped their fellow nature conservationists establish monitoring schemes that in time will give them the information to tell their governments about the changes that must be made to conserve their countries' birds.

A good example is the new monitoring scheme set up by the BirdLife International partner in Spain, the Sociedad Espanola de Ornitologia (SEO). While the RSPB helped to establish the scheme, our Spanish colleagues didn't need an awful lot of help beyond encouragement, moral support and a bit of funding because they were, themselves, so skilful and proficient. But I remember standing with David Gibbons in the rolling arable plains south of Madrid looking into the skies trying to work out how many larks were singing and how many of them were short-toed, Calandra, crested or thekla.

The Spanish common bird monitoring scheme was established in 1996 and has now documented rises and falls in many Spanish bird populations. I notice that the report on 2008's birds has an introduction from the Deputy Director for Biodiversity in the Spanish Ministry of the Environment, which just goes to show how NGOs can capture government attention armed with some hard facts. Similar engagement has occurred, of course, in the UK but also in many other European countries where bird monitoring is becoming better and better established. Once everyone is counting their birds it becomes much easier to produce a continent-wide picture of trends, see the big picture more clearly, and examine the national variations around that big picture.

Conservation priorities – the cube

OK – so let's imagine that bird monitoring is perfectly accurate, covers all species, and the results become immediately and freely available. What do you do with this information?

Deciding where to spend conservation resources is an art and not a science. Although a bit of logic probably doesn't do any harm.

Back in the 1970s the RSPB acted as though its job was to look after rare species such as avocets, ospreys and marsh harriers. And we (long before my time) did a pretty good job in conserving those species. But to my mind, and certainly in the modern age, the challenges of nature conservation are more to do with keeping common species common than in helping rare species.

There are several types of rarity though. There are species that are rare because they naturally occupy small geographic areas (like those that live, and always have, at the top of a single mountain) or rare habitats (such as coral reefs). There are those that are rare because they occupy the top of a food chain (peregrines tend to be rarer than pigeons), and those that are rare here but common elsewhere. And there are those that are rare because although they used to be common, something nasty has happened to them (the prime example being the passenger pigeon which went from being the commonest bird in the world to extinction in only about 50 years).

In the past, we probably spent rather too much time protecting those species that were rare in the UK even though they were common elsewhere, at the expense of common species which were in trouble. But that's where great monitoring schemes come in – ideally a national conservationist would know all about their own country's wildlife, and other countries' too. And then make rational decisions. But what are rational decisions? What exactly are we trying to achieve through nature conservation? A question for a few friends and a few bottles of wine, I would say.

The man in the street might say that preventing extinctions is what nature conservation is all about – but I believe he would be wrong. He would have got this idea from all the publicity and talk about how the tiger (or giant panda, blue whale, etc.) will soon be no more. Now, don't get me wrong, I don't want the tiger to become extinct. But extinction is the last gasp in a long story. Indeed, I don't want tigers to be anywhere near extinction so protecting the last few hundred is better than nothing, but is still a pretty poor show. But the tiger is a good example in that it is a widespread species whose numbers have probably fallen from around a million about a century ago, to fewer than 10,000 today (a decline of some 99%). It's the 1% loss somewhere around the middle of that decline that gives me the most pain – the last 1% is sad but maybe not the most important. So if we stop the extinction of the, say, Azores bullfinch, but allow the skylark to decline to half its recent numbers, then does that represent winning or losing? It's your choice – or similar types of choices are yours if you are a senior member of a conservation organisation.

My personal conclusion was that we need to take not only a global perspective (which means that the occasional nesting wood sandpiper is not the highest priority for UK conservation) but that we also need to treat rapid declines of populations as important even if the species declining are still relatively common.

All this came to a head when a group of us, comprising staff from the Nature Conservancy Council (NCC), BTO and RSPB, started a revision of the Red List of British birds. Our predecessors, including the incomparable Colin Bibby, had done the difficult job some years earlier in producing a book on UK bird conservation priorities. They had wrestled with the 'rare here/common there' conundrum and with the 'it's declining but it's *still* common' issue. They had also added the important thought that the 'common here/rare there' species were important too. All those seabirds that were actually increasing in numbers, but whose global populations were concentrated in the UK were surely important as well?

The existing Red List consisted of 117 species out of the 500+ which occur in the UK in some sense or other. Its compilation was a great achievement – but the list did include the cirl bunting (a formerly widespread species which had declined dramatically) and the Lapland bunting (a rare species on the edge of its range in the UK), as well as the greenshank (because everyone thought it was threatened and couldn't bear to leave it out) and the gannet (because of the global significance of our populations). It did not contain declining farmland species, partly because the data on them were not as clear and not as easily available as they are now, and partly because some argued against the importance of including widespread declining species.

Let me be clear, the Red Data List produced by Colin Bibby's team was a massive intellectual breakthrough. In our own little world it was an unsettling and exciting revolution in thinking. They kicked off a process that subsequently developed its own inevitable momentum, and it fell to a group chaired by me to capitalise on the uncertainty created by these revolutionaries.

Now, I believe that every group which ever presides over a process like the Red Data List review tells itself, even promises itself, that it is just going to tidy things up and not meddle too much because everything is basically alright. We said that, we promised that, and then we ignored it. The talented group who worked with me included David Gibbons (then of the BTO), Tom Tew (then of the NCC), Gwyn Williams and Richard Porter (of the RSPB) and Graham Tucker (of BirdLife International).

Our contribution was to introduce a true traffic light list of green, amber and red species which allowed us to use the amber category for those species of high international importance (but not at the moment in trouble), rare and highly localised species (but not declining) and species whose declines were moderate. We kept the Red List for those species in steepest decline (more than 50% lost over the past 25 years) and those few species which are listed as globally threatened.

This new(ish) way of looking at conservation priorities was written up as a paper in *Ibis* and presented at a BOU conference held not far from The Lodge in spring 1994. We called it the conservation cube as we used three axes – decline, international importance and scarcity. I remember using all my children's *Duplo* to produce a coloured cube as a visual aid at the conference.

Just as the Red Data book had stirred up a bit of disquiet as it presented a new way of looking at things, so did our conservation cube, although to a lesser a degree. Looking back, our elaboration of the subject generally seemed to help people develop a mental framework that they found helpful. And because of that, although the conservation cube idea has been sensibly tweaked over the years, largely due to work by David Gibbons, it provides a basis for setting bird conservation priorities to this day. And its influence has spread into other taxa.

So, while our conservation cube approach caused a few ripples initially, after we had assessed all UK species against its criteria and come up with the new Red, Amber and Green classifications for all UK birds, it caused a tidal wave of interest. This is quite usual too – it's easy for people to agree with a logical argument but sometimes they then moan at its consequences when they become manifest – although there wasn't that much moaning. The tidal wave of interest came from the media.

The conservation cube idea appeared after the BTO had brought out its book on population trends describing the cumulative results of the CBC, and suddenly the Red List was flooded with common farmland birds. The Red Data book contained a list of possible candidates for the next list which included a number of farmland species, but they were just suggestions and were scattered amongst a range of other species. When we brought out a revised bird Red List we pointed out that it was littered with common but rapidly declining species such as song thrush, skylark, corn bunting, linnet, tree sparrow, grey partridge and turtle dove. The message was new and clear – if you look at UK bird populations, the place where most is going wrong is the farmland through which you drive or in which you live.

All the papers covered the news – a list is too easy a story to cover for any journalist to ignore! Most papers gave it a full page with pictures. I was on the early morning BBC Radio 4 *Today* programme, talking to John Humphreys (for the first of many times to come) and we also featured on Radios 1 and 2. TV cameras appeared at The Lodge and the state of the UK's birds was a headline story all day. And then we got back to normal life once more!

Over the last 18 years, the composition of the Red, Amber and Green lists has changed a bit but not as dramatically as in that reappraisal of the early 1990s. Woodland birds have edged more into the Red List – species such as lesser spotted woodpecker, redpoll and hawfinch. It has also become clear that quite a lot of trans-Saharan migrants are in trouble, but otherwise the Red List is still dominated by farmland species that were much more common not that long ago. And the Amber List mainly consists of internationally important seabirds and waterfowl, and a number of rare and localised species.

Doing something – moving up the curve

Once you have your monitoring in place, and a way of thinking about the results it produces so that you prioritise your actions, you must get on with trying to do some good in the world – how about saving some species through conservation action?

But what to do? We think of there being a conservation tool box – and in fixing any threatened species, it is important to choose the right tools. So what's in the tool box?

Admitting that you don't know what the problem is can sometimes be the first step. Research to identify the causes of decline, or more importantly remedies for the decline, is often the first step on the road to recovery.

Few species are the subject of people's deliberate nastiness these days – legal protection is in place for all birds to some extent, and for most species to a good extent. These battles were won back in the distant past – by the founders of the RSPB who won protection for birds being killed for their feathers – and then more comprehensively by the Protection of Birds Act of 1954. There are some species, like the hen harrier, where deliberate persecution is at the top of their problems (see Chapter 11) but most are threatened accidentally rather than deliberately. Species where legal protection, or the proper enforcement of existing legislation, is needed include a few raptors such as golden eagles and hen harriers.

Site protection (see Chapter 6) is a useful tool for those species which occur in only a few places or where their populations are very clumped. This might entail notification under domestic legislation as a SSSI (or ASSI in Northern Ireland – Site/Area of Special Scientific Interest) or under EU legislation as an SPA or maybe a SAC (Special Area of Conservation). These designations impose constraints on how the landowner can use their property – but the restrictions are not usually that onerous. However, if the site is owned by a nature conservation organisation then it can be managed specifically with its conservation interest in mind. The RSPB now manages over 140,000 ha of land in the UK as nature reserves (see Chapter 9) – an area a little larger than Bedfordshire.

Direct conservation action is sometimes needed for particular species. Reintroduction projects, such as those described in Chapter 8, can be useful where it is clear that a species won't get back to suitable habitat under its own steam very quickly without a helping hand. Guarding rare species from disturbance or from egg collectors are other types of direct action.

Some species need widespread remedies – those farmland birds are affected by what thousands of land managers do to the land. Persuading farmers to manage their land in wildlife-friendly ways can sometimes be done face-to-face when an individual is sympathetic to wildlife, but sometimes the most effective route is to influence landowner behaviour through government payments or regulations (see Chapter 7).

Some of the things that affect birds really do need to be fixed at a policy level – climate change, over-fishing, water pollution are all really big issues that can potentially be tackled at the stroke of a pen – but it's getting to the person holding the pen that takes the time (see Chapters 10, 12 and 13)!

Choosing the right tools at the right time, and deploying them with the right amount of force and consummate skill will enable species to increase in numbers. The RSPB has quite a few successes of which it can be proud, and also quite a lot of unfinished jobs and a few remaining wonky shelves!

Throughout this book I will touch on all the tools in the conservation tool box, and at the end suggest which ones are most useful and which need sharpening.

Lessons learned

Although counting birds might look almost impossibly daunting in theory, the birding and conservation communities have got there in practice. There are tweaks that could be made to the current system which would make things a bit better and a bit quicker but British bird monitoring is rightly the envy of many others. And it's not a luxury – you need to know what's happening and what's working if conservation action is to be targeted effectively.

Conservation of other taxa is compromised because there isn't anything like the same level of information on species population trends – plant conservationists might be the ones leading the discussions about how intensive farming displaces wildlife if only botanists had got their act together a few decades ago – after all, plants just sit around waiting to be counted!

You do need a system for deciding what *not* to do – unfortunately conservation needs exceed conservation resources, so it is right to put it that way. The conservation cube framework isn't the only way to make decisions, but it's not a bad one given that it takes into account the wider context of a species global status and whether it is in trouble in the UK. I am particularly proud of the fact that the system which I helped to establish treats declining common species as important conservation subjects and helped nature conservation practice in the RSPB and elsewhere to move away from the rare and the localised.

But then what you need to be able to do is to pick the right conservation tool from the toolkit – will nature reserves help or do you need to engineer policy changes in government? Can you reintroduce a species or do you have to chat to lots of farmers about how they go about their business? For some species you need to combine several approaches and you need to be able to plan how to allocate your resources. Bird conservation has led the way in these debates and its influence has permeated the thinking of governments and of other NGOs.

There is also a general lesson that comes back to me loud and clear whilst thinking about these essentially internal, domestic conservation-specific changes that we made to monitoring, and choosing priorities and appropriate actions – and that is that people don't like change. The change from the CBC to the BBS was, in retrospect, a much-needed and obvious one and yet there were heels dug in all over the place. I'd say that the same things happened every time that a slightly different way of addressing conservation priorities was invented – lots of people don't want to see change. And yet now, no-one is campaigning to ditch the BBS and go back to the CBC and it is widely recognised that the BBS is a much better scheme all round. And that doesn't mean to say that the BBS is perfect, nor that the CBC was rubbish – each is far from true – but it does mean that progress has been made.

I guess that throughout my life I have been a radical – eager for change for the better – and that has applied across the board in the way that I have looked at science, nature conservation, organisational management and politics. It may seem odd, in what might be the driest chapter of this book, to end with a paragraph in praise of radicalism but it seems to me that I learned a lot from working with others to change the way (just a little bit) that we think about nature conservation. I learned that you just have to push ahead if you feel change is needed and take as many people with you as possible – but it won't be everybody. And, however determined the opposition to change might be, once you have implemented that change the fuss dies down and rarely do people want to go back to the *status quo ante*. Just as, in my little world, hardly anyone would want to go back to the CBC or the muddled system of assigning conservation resources prevalent during my early days at the RSPB, so no-one campaigns for the return of lead to transport fuels or the removal of seat belts from cars, and on a higher plane, no-one argues for the reintroduction of slavery or the de-emancipation of women. Progress is marked by opposition until it's achieved, followed by grudging acceptance, and then finally, by complete reconciliation. It's getting the change made that is often difficult, not keeping it afterwards.

> *A man's interest in a single bluebird is worth more than a complete but dry list of the fauna and flora of a town.*
> **Henry David Thoreau**

Chapter 5

Is it ever right to be nasty to birds?

Every creature is better alive than dead, men and moose and pine trees, and he who understands it aright will rather preserve its life than destroy it.
Henry David Thoreau

Most people working in nature conservation actually like nature. Most of us rarely think very hard about why we are attracted to the natural world, but we are. We often talk about the wonder of evolution, the beauty of nature and the importance of the natural world in sustaining our own existence on this planet, but when you show an image of a lapwing chick to an audience and hear a collective 'Ahhhh' then you know that you are tapping into something deep and emotional, rather than deep and intellectual.

It's striking that the support for different natural taxa varies so considerably in line with their cuteness. Popular support for mammal and bird conservation is way higher than for insects or plants, and that can't be anything to do with the relative evolutionary perfection of these groups nor their ecological importance – it must be to do with how cute they are and how much we feel that they are a bit like us. When David Attenborough shows a butterfly emerging from a chrysalis we go 'How lovely' but when he shows a wildebeest calf getting to its feet for the first time, we go 'Ahhhh' – don't tell me that it isn't the cute factor at play. We empathise with the wildebeest and its mother in a way we cannot with a butterfly.

It's the birds and mammals that we like the most. Man is a strange mammal so it's not surprising that we feel kinship and closeness with birds as well as our fellow mammals. Whereas most mammals live in a world of smells and touch most birds live in a world of colour and sound. Most mammals are nocturnal (all those bats and rodents) whereas most birds share the daylight with us. And although our lives are more governed by smells and pheromones than we actually recognise, you will much more often find a group of human mammals talking about that good-looking man or woman rather than that good-smelling one.

Birds and mammals both have 'babies' and provide some form of parental care. The young of even the most precocial of gamebirds and waders are shepherded by a parent who leads the brood in search of food and looks out for danger – and who is often prepared to protect its young by fighting off predators or apparently risking its own life by feigning injury to distract attention towards itself. All that suckling and grooming and cuddling looks pretty similar to the way that our own young are reared but the whole idea of a blackbird or an eagle bringing food back to the nest also reminds us of our own parental duties and responsibilities.

And the cute-factor really does matter. As I write these words I have the autumn magazines of two of my favourite nature conservation organisations in front of me. Butterfly Conservation has a red admiral on its cover, and a very fine looking beast it is too, but the Wildfowl and Wetlands Trust magazine has a spoon-billed sandpiper chick looking out at me and there really is no contest. And the WWT marketing staff knew exactly what they were doing when they placed the words 'Bringing up baby' in bold letters underneath the image. Those working on other taxa sometimes speak disparagingly of all the money that goes into charismatic furry and feathery species, but it is hardly surprising that that's where the money goes rather than into the slugs and liverworts.

Our language is full of examples which show how close we feel to birds. For example, a 'cool chick' is very different from a 'cold fish' and you probably wouldn't want to be known as a reptilian character. All that noise that birds make to attract mates and defend their territories (sex and violence, in fact) we call song and do not feel any compunction in so doing as it seems so similar to those noises that we make for somewhat (but not entirely?) different reasons. Young women are 'birds' who may turn into 'nest builders' and a man may act in a 'cocky' way.

Birds and mammals make it easy for us to anthropomorphise, in ways that redwoods, starfish and spiders do not. A 'bunny-hugger' may be a bit soft but a 'tree-hugger' is a bit soft in the head too. It's easy for us to love mammals and birds because they are so like us. And that love becomes apparent at a very early age. Most children seem to like nature even if it manifests as their wanting to keep a pet – their love of nature often starting with the love of an individual animal.

Feeling affection for individual animals predisposes us to care about their populations and the fate of life on Earth, so it's no great surprise that the RSPB was founded when a group of women started to campaign against the killing of birds for the millinery trade. They believed that birds were beautiful in their own right – and far more beautiful flying around the world than dead and poking out of somebody's smart hat.

The scale of the slaughter of birds such as egrets (for their head plumes), roseate terns (for their tail streamers), ospreys (for their crown feathers), etc. was so great that it was believed to be affecting the population levels of those species. And, it just seemed so clearly wrong anyway.

Given our deep visceral feelings for individual animals, and that they often lead us into our love of nature conservation, it is not surprising that some of the most difficult and emotional issues faced by conservation bodies are those rare examples where conservation ends must be met by killing animals. In this chapter I look at several such cases, and there has been no better example in recent decades than the RSPB's support for a cull of ruddy ducks.

Ruddy ducks

Having spent most of my school holidays cycling around Chew Valley Lake watching birds, I was familiar with the ruddy duck – an attractive stifftail duck seen at the lake but not many places elsewhere. This species is native to North America and had escaped from the WWT centre at Slimbridge, up the road in Gloucestershire.

This attractive bird even featured in my copy of the Peterson field guide where it sat next to a bird never seen in the UK – the white-headed duck native to Mediterranean wetlands such as the Coto Donana – seemingly a million miles from the Bristol Waterworks' reservoir that was my local wetland haunt. And there's the rub, the escaped North American ruddy ducks (yes, oversexed and over here) increased in numbers until there were thousands in the UK and they were one of our most successful exports – spreading to continental Europe and finding their way to their native European congeners.

When two closely related species meet often nothing happens, but in this case it did. Not only did the randy Americans mate with their hot-blooded Spanish relatives (and obviously *vice versa* – it takes two to tango) but these couplings produced viable young which were themselves interfertile with their parent species. So, if UK ruddy duck numbers continued to increase (as they were doing), and if they continued to colonise continental Europe (as seemed inevitable) it looked as though the growing numbers of ruddy ducks would increasingly threaten the existence of Europe's only stifftail duck. It could be a little like the grey squirrel/red squirrel story all over again except this time one would end up with greyish-reddish squirrel equivalents everywhere and the deed would be done through sex rather than disease. To make things worse, the white-headed duck was ill-prepared for this assault, already being listed as a globally threatened species because of hunting pressure and habitat loss and degradation. So we were dealing with a potential extinction here – and the stakes were correspondingly high. Little did I know when I enjoyed the sight of those cute ruddy ducks at Chew Valley Lake in the 1970s that I would be in favour of their eradication in the 1990s.

The great Sir Peter Scott, one of the finest naturalists of the 20th century, was in some ways responsible for all this since the ruddy ducks had escaped from Slimbridge. According to the 'polluter pays' principle, the WWT should be paying for the ruddy duck eradication, but instead the EU taxpayer is. At the time of writing the eradication programme is going very badly for the ruddy duck, very well for the white-headed duck, and you, the taxpayer are getting good value for money from it, in that UK ruddy ducks have been cut in numbers from over 6,000 to fewer than 200 individuals. It's those last 200 that make all the difference though – if they aren't killed then the problem will re-emerge in the future and need to be tackled again. Even so, there is always the risk that ruddy ducks might escape again from captivity or just possibly re-colonise the UK from smaller continental populations that derived from a mixture of escapes and British exports.

I was fully involved with the RSPB's position on ruddy ducks and I can remember how difficult a decision it was. It wasn't easy to make – it wasn't universally agreed amongst staff – it wasn't the least bit popular with our membership – it led to death threats to Graham Wynne and myself – the occasional bouts of media coverage were difficult – and at heart most nature lovers found killing wild creatures a bit unpalatable. If I had to make the decision on whether to cull ruddy ducks a hundred times again in the future (and it wasn't mainly my decision to make at the time) then I'd probably make the same decision eight times out of ten. The other twenty percent, I may have been swayed. Sometimes nature conservation requires difficult decisions to be made and someone has to make them. I am proud that, rightly or wrongly (and I think rightly in this case) the RSPB chose the more difficult and uncomfortable route on sound conservation grounds.

Of course, the ruddy duck is an introduced species in the UK, and indeed in Europe. Introduced species often cause conservation and economic problems when they are inserted into the native fauna and flora. Across the world, species introductions are a major cause of extinctions. But that doesn't mean that every introduced species causes havoc – many are, in conservation terms, benign additions to the native wildlife. And there's the rub, if all introduced species caused massive problems then there would already be much better regulation of transporting species around the globe and every country would have plans in place to nip the possibility of non-native species becoming established in the bud. But we shy away from taking drastic action that involves killing introduced species because of the expense or because of legitimate moral scruples over condemning a species before it's been found to be a problem. And we have been far too careless about species introductions far too often in the past.

Many of the past introductions, deliberate or accidental (but predictable) can be laid at the door of the landed gentry as only they had the land, money, leisure time and resources to carry out such projects. We have the former Dukes of Bedford to thank for the millions of pounds of damage caused annually by muntjac deer, the Brocklehurst family of Cheshire for the introduction of grey squirrels into England and the subsequent loss of our native red squirrel, and none other than the royal family, in the shape of King James II, to thank for the Canada geese now ruling our parks. The 'polluter pays' bill for these introductions would each amount to many millions of pounds.

Predator control

Conservationists can often persuade themselves of the need to be nasty to introduced species when they are being nasty to our native wildlife. It feels like one is stepping in, albeit with qualms, to redress an unnatural situation caused by human activities. So, for ruddy ducks it is a respectable argument to say that 'ruddy ducks belong in North America, they are not native to Europe and they are causing problems for white-headed ducks that can't be solved in any other ways as far as we can see – so let's kill them'. But when a fox eats a lapwing chick it looks like a natural event. Whose side should we be on? Lapwings and foxes have rubbed along together for thousands and millions of years and so who are we to interpose ourselves between the fluffy lapwing chick and the sharp teeth of the fox? And nobody ever thinks to interpose themselves between the wetland insect larvae and the sharp beak of the lapwing chick.

Conservation organisations are a bit coy about the amount of predator control they do as it can never be enough for some critics and can never be too little for others. At the RSPB we developed a policy that wouldn't please everyone but was designed to be appropriate for a nature conservation organisation that is clearly on the side of nature, and not just of cuddly nature.

The RSPB generally uses predator control as a last resort and not as a standard management prescription. Ideally there would be no shooting of foxes or trapping of crows at RSPB nature reserves, but where predator numbers are shown to be putting species of conservation concern at risk then we do instigate predator control. Most is carried out to protect ground-nesting wading birds, such as lapwings, although capercaillies, terns and cranes are also species which may have benefited.

Over the years I became more and more convinced that the RSPB needed to do a bit more predator control – but not nearly as much as some of our external critics would have us do. Foxes and crows are wildlife too and predator control costs money. And, we shouldn't forget that, basically, killing things is a morally dubious activity. However, sometimes nature needs a helping hand against other nature, for a while at least, and then I feel we should be prepared to act.

The keenest proponents of large scale predator control are those who either practise or are supporters of field sports. Visit the Game Fair one July day and you can have a hundred conversations about how we need to rebalance nature, how nature is wonderful but needs management to survive, and a whole lot more prejudiced drivel. It is clear that if your sport and enjoyment depends on the number of pheasants you shoot then anything that eats your pheasant before you can shoot it is an enemy. Foxes and crows are in competition with the pheasant shooter and the pheasant shooter is paying for his enjoyment and therefore wants as much enjoyment as possible. So a ruthless approach to removing the competition is entirely logical from the shooter's point of view. It therefore becomes convenient to exaggerate the need for predator control because otherwise there is a danger that your sport just looks like a bloodbath of pheasant shooting dependent on a bloodbath of fox and crow killing – and who could hold their head up if that were how they got their kicks?

I know very few of the keenest proponents of legal predator control who aren't also quite keen proponents of more control of birds of prey and other currently protected predators. There are thin ends of larger wedges at play here.

So my line on predator control would always be that it cannot be entirely ruled out, that it is sometimes necessary to achieve conservation goals, but that it always should be considered very carefully and only employed as a last resort, and only then when it is thought that it will actually make a difference.

Pests

Just as a weed is a plant in the wrong place then some birds can get themselves into trouble with other users of the countryside. There is something which resembles a pest list enshrined in our legislation – there are some birds which can be killed by anyone under something called the General Licence. This is a licence to kill but there are some, although not many, constraints. A landowner can kill certain species of bird without a licence for reasons of protecting livestock, human health, property, wildlife, or air safety. The species which can be killed for these reasons are (in England – details differ slightly elsewhere in the UK): carrion crow, rook, jackdaw, magpie, jay, wood pigeon, collared dove, feral pigeon, Canada goose, ruddy duck, herring gull, lesser black-backed gull and great black-backed gull. Not a daft list – although until recently, and in other parts of the UK, the house sparrow and starling were or are still listed despite their declines in numbers. It took a fair amount of lobbying to get house sparrow and starling removed.

You'll notice that protecting game is not an allowable reason for using the General Licence – so it is just arguable, and no one has tested it, that all gamekeeping is illegal in the UK. I've always wondered whether this is a reason that game shooters go on at such length, with differing degrees of persuasiveness, about the wildlife benefits of their management practices.

Egg collecting

I was born a little after the Protection of Birds Act made egg-collecting illegal and so I never had the training of searching for nests in the hedgerows and learning field-craft in that way. But then again – I never wanted to so the law was irrelevant! However, many of the expert ornithologists to whom I look up, had their own egg collections as boys and learned huge amounts from them. A few boys taking a few eggs from a few blackbird nests probably didn't do much harm but when this past-time turns to the type of obsession that only we blokes can create then it becomes a conservation issue – and, I think, a mental health issue.

Birds' eggs are beautiful objects and I love the look of them, however, a bit like the campaigning Duchess of Portland I believe that such beauty is better enjoyed through knowing that its owner is alive and well rather than dead and, in this case, sitting in a cabinet with a label on it.

But despite its illegality (some might say partly because of it) egg-collecting continues in this country. It would make a fascinating sociological study. Why are almost all egg-collectors men? How did they get into their 'hobby'? What does it do to their relationships with their partners? Do they break other laws more often as a result of getting inured to being a criminal? What are the contacts between the proponents of this obsession? Is this a particularly British trait? Are they a particular personality type? Do they go birdwatching? Do particular egg-collectors specialise in particular species, and if so why? Do they have small penises?

Whatever their character traits and behaviour, they are definitely still out there. Every now and again huge egg collections are discovered comprising thousands of eggs, which have been assembled by obsessive men, mostly through their own diligence and in their own lifetimes.

In 2010 when a pair of red-backed shrikes nested on Dartmoor, egg-collectors were sniffing around the nest site and were chased off by volunteers guarding the nest. Without the type of expensive round-the-clock protection which is often organised by the RSPB, rare isolated birds would fall victim much more often to the hand reaching into the nest and carefully removing the eggs, before blowing their contents and privately delighting in their hollow beauty.

We may never know all the details of this murky world but the fact that egg-collecting still goes on is odd and somewhat disturbing when one realises the scale of its toll on some species of conservation concern. The RSPB Investigations staff are an odd but admirable bunch of folk and every now and again they receive evidence which leads to the conviction of some long term egg collector. In recent years, cases have included single collections involving thousands of eggs of protected species, and some egg collectors have been arrested and convicted multiple times. Collections have been found hidden under floorboards or in secret rooms and are sometimes accompanied by meticulous falsified notes which attempt to show they were collected before egg collecting was made illegal. It's a strange world, a man's world, and a strange man's world.

Egg-collecting from the nests of rare birds falls firmly in the category of harmful enjoyment – there is no greater justification for it. We do not need to control the number of red-backed shrikes in the UK by preventing them from nesting – there is no wider public benefit. And the people carrying it out clearly enjoy the whole thing. No-one forces them to do it – they just want to do it.

Game-shooting

It's a case of the good, the not so good and the ugly, in my view. The RSPB's Royal Charter, established over a century ago and tweaked a few times since, although not in this respect, prevents the RSPB taking a view on legitimate field sports except where they affect the Objects of the Society. The RSPB has always interpreted this as meaning that it can't say that shooting birds is wrong but it can comment on the way that shooting is carried out (e.g. the use of lead ammunition, cold weather bans, shooting seasons) and that it has free rein to comment on raptors being killed illegally to increase game bags.

I'll deal with raptor killing elsewhere in this book, so let's consider now all those birds that are shot for sport. Apart from the Royal Charter, why should the RSPB be reasonably relaxed about this going on? Surely a bird conservation organisation should be against the killing of birds? End of?

I totally understand why some people feel that the killing of wild creatures for sport, or even for food, is wrong. I'm neither vegan nor vegetarian myself but I do care about animals and so I don't feel that hunting wild birds is something that I would want to do. But for me it is a 'live and let live' subject – rather ironically you may feel. If those who hunt do so in a sensible way then as far as I'm concerned we can rub along together. As I say, for me it's live and let live.

But there are, even for me, gradations of indifference to hunting. Wildfowling seems to me to be the 'good' end of the spectrum – more akin to harvesting a natural resource than anything else. The main wildfowling species – mallard, teal, wigeon and a few geese –all enjoy pretty healthy populations and wildfowlers are, in the main (like birders), a generally responsible bunch. It's a pretty sustainable system all things considered.

Some forms of partridge and pheasant shooting are quite sustainable too

– although here, to my mind, we are moving into the 'not so good' area of shooting. Some shoots operate on the basis of taking 'wild' birds – but in most cases birds are reared and released every year. In total, more than 30 million pheasants are released into the UK countryside every year for sport shooting. And I find it difficult to believe that such huge numbers of this large and non-native species have no impact on the ecology of the countryside. I guess it's a matter of balance between habitat management for pheasants, which benefits other wildlife (by providing cover crops, woodland management and winter feeding), and the impact of all those pheasants on predator numbers (many of them end up feeding foxes and crows) and more natural food supplies (they are all gobbling seeds a lot of the time).

Then there's the downright 'ugly'. Grouse shooting is a sport greatly misunderstood by the lay person. I think that if the woman on the Clapham omnibus ever thinks of grouse shooting then she imagines it as a rather natural pastime where birds live in peace before some rich and slightly oddly dressed toffs turn up for a day out to shoot a few. This picture is very far from the truth.

Grouse shooting is a high investment business where the returns are a bit unpredictable. From the opening of the season on the Glorious Twelfth (of August) a day's grouse shooting involves the guns standing in a line of butts as the beaters drive the grouse towards them. Grouse fly fast and pack together, so I can see that it is indeed quite thrilling and skilful to drop them from the sky. But let's get away from the idea that this is a harvest of a natural population. Red grouse are, unlike pheasants, native to the UK, and indeed our red grouse is the British race of the continental willow grouse. But the densities of red grouse needed to make grouse shooting profitable in the way that it is currently practised in the UK, are way in excess of what could be regarded as natural densities – maybe as much as a hundred-fold higher. These phenomenal densities are achieved through a variety of means but necessitate much of upland Britain being given over to the requirements of a single species just so that it can be shot later in the year.

Red grouse depend on heather for nesting and for food, and they do best when there is a mixture of short heather for feeding and long heather for nesting. To provide this intricate patchwork, much of northern England and south and east Scotland is burned, very carefully, on a 4–10 yearly rotation. A zero tolerance policy is adopted regarding predators in the run-up to the breeding season and so foxes, stoats, weasels, crows, magpies etc., those species which can be killed legally, will be ruthlessly disposed of. The high densities of grouse on a grouse moor render them susceptible to various parasites and diseases and so piles of medicated grit are provided for the birds to peck at, and on some moors gamekeepers will catch birds at night and douse them with medicines. On top of this, protected birds of prey are too often illegally killed – but I do mean to keep that until a later chapter (funny how it keeps coming up!). As you can see, the pack of grouse pushed over the guns in late August is the product of a lot of work in the preceding months – and the creation of an environment that leaves no room for either predators or other bird species with differing needs

Game shooting is a sport, and people pay serious money to take part in it, so we can assume that this wildlife-killing gives pleasure to those who practise it. I can see that it would. Talking to game shooters it is the pleasure of being out there in the countryside, the pleasure of seeing wildlife, the pleasure of an accurate shot cleanly taken. I can understand all those things and I can see that they all come together in, say, wildfowling. Shooting clay pigeons in a dull field surely cannot compare with shooting wigeon at dawn on an Essex saltmarsh. I can begin to understand all that, although not to the point of wanting to participate myself. But the big pheasant shoots where each gun may bring down over a hundred birds in a day seem so close to ritualised killing that I struggle to understand where the enjoyment can be found. Yes, the day is spent in more attractive surroundings than a clay pigeon shoot, but the distortion of land management needed to breed and release such huge numbers of a non-native species so that they can be driven towards a line and gunned down, hardly provides a splendid aesthetic? There is indeed skill in the killing but the large numbers of deaths involved speak to me of a lack of sensitivity about nature and killing which troubles me. So much of the enjoyment seems to be related to the sheer numbers killed and I can't empathise with a pastime that delights in such slaughter.

In contrast, the wildfowler, whose club owns, protects and preserves the wetland where he shoots, takes his few ducks home for a meal for the family and seems a long way from the highly managed pheasant or grouse hunter shooting on land managed entirely to provide big bird days. When an apologist points out that grouse, pheasants and ducks are all edible I am led to reflect that we aren't going to starve for want of a little game meat treat.

Conclusions

Killing wildlife is bound to be a tricky issue for a wildlife conservation organisation – and it is. It's an emotional issue, and so it should be because, generally speaking, killing things is wrong – or it is at least a large enough moral issue as to be questionable. During my years at the RSPB I spent rather too much time dealing with debates about killing things. In any controversial issue it's important to get your thinking straight – it doesn't mean that people will agree with you, but it does mean that they can't find contradictions in your behaviour and criticise those.

I find it useful to think of these issues along two axes of thought. The first is to do with the necessity of killing a bird – is there a reason for it? Does it lead to a better world or a worse one? And the second is to do with the enjoyment that a person derives from killing a bird – is it done with enthusiasm or regret?

Generally speaking I have a problem with forms of killing that involve lots of pleasure and rather little necessity – so that means I probably won't be taking up grouse shooting as a hobby. But don't get me wrong – I wouldn't want to ban the legal pursuit of these pastimes. It's just that a person goes down in my estimation a little if they derive pleasure from killing things unnecessarily.

On the other hand, I can live with myself being in favour of ruddy duck control because there doesn't seem to be an alternative solution to the conservation problem they pose. I gain no pleasure from it – I just regard it as a necessary evil. Sometimes people have asked me whether I would pull the trigger to kill a ruddy duck myself and the answer is that I would – but I wouldn't enjoy it.

The same two axes can be helpful in thinking about other forms of killing – wars are justified when they prevent acts of evil and when carried out with regret, but are indefensible when enthusiastically waged for personal gain. And killing cows is acceptable, one can argue, for the necessity of eating (although a vegan would take another view) but would be seen as sickening if carried out solely for pleasure.

> *The assumption that animals are without rights, and the illusion that our treatment of them has no moral significance, is a positively outrageous example of Western crudity and barbarity. Universal compassion is the only guarantee of morality.*
> **Arthur Schopenhauer**

CHAPTER 6

Special places

Everybody needs beauty as well as bread, places to play in and pray in, where nature may heal and give strength to body and soul alike.
John Muir

It is a universally acknowledged fact that some places are better for nature than others. When I go out birdwatching I usually go somewhere specific rather than just anywhere. I'll go to the gravel pits near my home because they attract more species, and larger numbers of some species, than other areas. And I won't just go to any gravel pit – I tend to go to some particular ones which have proved better for wildlife than others.

The theory

At a global level, Colin Bibby and BirdLife International showed that you only need c. 5% of the world's land surface to include the entire global distribution of 25% of the world's birds – provided you find the right 5%.

So natural wealth is a bit like financial wealth, it tends to be found in big piles in some places and pretty sparsely in others. It was ever thus, and a consequence of nature's uneven distribution is that if you want to protect nature, then protecting the best places for nature is a very good start.

It's such a good idea that most countries have some system whereby the state recognises that certain areas are more important for nature than others. The National Parks across the world, whether Yellowstone in the USA or the Masai Mara in Kenya, Rathambore in India or the Great Barrier Reef in Australia, are recognised globally as being fantastically special for wildlife and are given legal protection in their respective countries.

Site designations should work for those species that are highly clumped in their distributions – those where a high proportion of their populations occur in a small number of places. These may be rare species such as bitterns, or the very localised species such as knot or gannet. Let's take the northern gannet as an example.

There are about 400,000 pairs of northern gannets in the world. Over 300,000 of these nest in western Europe (the others in Canada) and over 250,000 are in the UK where they are found at about 30 sites. Two sites, St Kilda and the RSPB nature reserve of Grassholm hold over a third of the UK population between them (yes, I have to admit that Grassholm is 'only' the second biggest!) so this is a very concentrated species in the breeding season. Let's just ram this argument home – Grassholm is an island less than 10 ha in area and holds about a tenth of the world population of this amazing cigar-shaped, plunge-diving, fish-catching, globe-trotting, guano-producing bird. So, snuggling up together, as they do, the whole world population of gannets could nest on an area of less than 100 ha, or one square kilometre. That is clumped!

And so we would hope that any system of site designation would have recognised the sites, particularly as they are traditional and long-standing, where gannets nest, and, phew, it does! Very few gannet eggs are incubated outside designated areas. The implementation of domestic and European Union (EU) legislation has done a good job for the gannet and indeed for many other seabirds such as kittiwakes, razorbills and guillemots, which seem positively expansive in their distributions compared with gannets, but which are also very concentrated in their distributions. And the designations also do a good job for species like knot, bar-tailed godwit, dunlin, grey plover etc. which concentrate on our estuaries on passage and during the winter.

Generally speaking, and without pre-empting future complaints about the details of the system, site designation is pretty good at recognising the best places for birds in the UK. It identifies key estuaries, seabird colonies, heathlands and wetlands very well. From nature's point of view, we have picked out the important sites nicely. We'll get on to how some people feel about this a bit later, but what, from nature's standpoint, are the remaining problems with the system?

One snag that might affect the gannet and a number of other seabirds is that their nesting colonies, as we have seen, are well covered while their real habitat, out at sea, receives scant protection. In theory, the site designation mechanisms should apply and work as well over water as over soil, but they don't. This is partly because the marine environment is fundamentally more variable than life on land, and physically and ecologically more fluid. Protected marine areas should move around over time to do as good a job at sea as protected areas do on land. We don't know quite enough about the marine environment to be sure of how they should be constituted. The RSPB is gathering lots of useful information on this question using data loggers to track seabird feeding movements and is building up a picture of which areas are most important and how mobile they are. But the fact is that successive UK governments have been pathetic at investing in studies of the marine environment and that means that the data just aren't good enough at the moment.

I've always felt uneasy about the use of protected areas in the uplands but maybe I worry too much, or take too scientific a view of the issue. Take the golden eagle as an example. There are far fewer golden eagles in the UK than gannets – only 500 pairs, and there would be a lot more if bad people didn't still kill them – so surely they are a species where protected areas could make a difference. But whereas the world's entire gannet population could nest within a single square kilometre, no golden eagle territory is that small. The range of those 500 eagle pairs covers perhaps almost a quarter of Scotland's land area – and the eagles are thinly smeared evenly over that area. I can see why the Scottish government has baulked at designating upland areas for eagles, harriers, golden plover and other species. But then again I have no problem whatsoever in seeing large areas of the Flow Country designated because of the importance of its moorland bird assemblages. I think it just shows that upland sites and species are at the extreme of what is sensible regarding protected areas.

The practice

In the UK we have a plethora of designations and labels to describe bits of land which might help to protect the nature that lives within them. There are National Parks, Areas of Outstanding Natural Beauty, National Nature Reserves, Sites of Special Scientific Interest (and in Northern Ireland, Areas of Special Scientific Interest), Ramsar sites, Special Protection Areas for birds, Special Areas of Conservation, local wildlife sites and others. It's a thick and murky alphabet soup.

All of these designations are important but, to make it simpler and not to lose too much of the truth of the matter, there are three designations, of two types, that are of predominant importance. These are the SSSIs (and ASSIs) which are based on UK (but now devolved) legislation stemming from the 1949 National Parks and Access to the Countryside Act, and then there are SPAs and SACs which are EU-wide designations stemming from two European pieces of legislation, the Birds Directive of 1979 (for SPAs) and the Habitat and Species Directive of 1992 (for SACs).

A whole book could be written about the history and evolution of site designations in the UK, and maybe someday someone will, but it won't be me and for this discussion we don't have to know all the details. But a few words about the origins of the national and EU-wide designations, how good their coverage is and what they actually mean in practice on the ground, will set the scene for the more interesting task of seeing how useful these designations are in the conservation toolbox.

Sites of Special Scientific Interest

The name SSSI is a turn-off. It doesn't trip off the tongue or come close to sounding like 'great place for nature' and I am quite sure that many keen naturalists don't have a clue where their nearest SSSIs are. The origins of the term go back to the post-war era and the 1949 National Parks and Access to the Countryside Act, but it was the 1981 Wildlife and Countryside Act that really created SSSIs as we now know them.

As well as biological SSSIs there are geological SSSIs. There are currently c. 4,100 SSSIs in England and they cover about 8% of England's land area.

SSSIs were never meant to encompass all the good places for nature in the countryside, but were designated to protect a representative selection of habitats and species. Designation as an SSSI doesn't mean that the site is preserved untouched for ever more, but it does mean that any activities likely to damage the site, whether they take place within it or nearby, and which are covered by the planning system have to be thought through very carefully before they are approved.

Natural England, the statutory nature conservation agency in England, has the lead role in designating SSSIs, identifying the types of activity which would harm each SSSI and has to be consulted on activities which might reduce its interest.

Although initially intended merely to protect examples of a wide range of good sites for nature, SSSIs have come to be seen, particularly in southern England, as the last refuges for important wildlife in the wider countryside. If you are enjoying a great day with nature you are probably in an SSSI whether you know it or not, and it's probably thanks to the legislation that protects them that you are having such a good time.

The Countryside and Rights of Way Act of 2000 (CROW) was a great leap forward in nature conservation. Part of the Act focussed on the management of SSSIs because existing legislation had been much better at preventing dramatic damage to them (such as their being concreted over) than it had at ensuring they were managed in ways that ensured that nature flourished within their designated boundaries.

The CROW Act established a system whereby Natural England assessed the condition of SSSIs (things are, of course, a bit different in other parts of the UK) and whether they were getting better or worse. The last Labour government even established targets for the condition of SSSIs – that 95% of them should be in favourable condition by 2010. This was never going to happen since at the time only about a third of SSSIs were regarded as being in good condition. However, and this is the good thing about setting targets, it generated a lot of positive action, not least that all these sites were properly examined. By December 2010 the Conservative government was able to announce, rather quietly actually, that 96.67% of English SSSIs (by area) were in a 'favourable' condition. In fact, 37.14% of SSSIs were regarded as being in favourable condition whereas 59.53% were in an unfavourable condition, but deemed to be recovering and therefore, for the purposes of this target, were regarded as already 'recovered'.

The CROW Act was a great achievement – it was secured following campaigns by a large consortium of conservation and countryside NGOs, and as well as improving the lot of wildlife it also secured a limited right to roam in England. The focus on the condition of SSSIs should be a lasting legacy for wildlife into the future. However, there will always be the temptation for government, and perhaps for its statutory agencies, to gild the lily and to claim that things are better than they really are. Even if we accept that 96.67% of England's SSSI area is either in good nick, or recovering towards it, nature conservationists should keep an eye on these figures over time. Given that 59.53% of SSSI area is still classed as unfavourable but recovering, and given that the recovery will often depend on concerted and prolonged good management (such as grazing of the right type, at the right times, by the right types of beast), and that given that the right course of management action may depend on continued funding during times when money is tight or when nature conservation is downgraded as a priority by government, then it is up to NGOs to be vigilant and to highlight any signs of back-sliding.

SPAs and SACs.

The Birds Directive came into force in 1979. One of its requirements was to designate sites, as Special Protection Areas for birds (SPAs) that would help to conserve the priority species listed on Annex 1 of the Directive plus any regularly occurring migratory species. As of February 2011 the UK is the EU state which has designated the smallest proportion of its land area (5.1%) as SPAs apart from the tiny Malta (4.9%). Most EU states have designated a proportion of their land area twice as large as this, many three times as large, some four times as large, and Cyprus and Slovakia have reached five times our percentage coverage.

Some of your favourite birding sites are SPAs – many estuaries including The Wash and the Humber, heaths such as the New Forest and the Dorset Heaths, and the North Pennines as well as a few reservoirs such as Abberton in Essex. Not surprisingly, given the importance of the UK's estuaries for migratory and wintering waterfowl, many of the sites are estuaries, and that is true of the French, German and Danish SPAs too – again rather unsurprisingly.

The Habitats Directive came into force in 1992 and achieved, in some ways, for the rest of nature what the Birds Directive had done for birds. It established Special Areas of Conservation (SACs) as its designated sites. The UK is bottom of the list of 27 EU states, most of which joined the EU much later on, as far as percentage of total land area designated is concerned. Whereas 5.4% of our terrestrial land area is designated as Sites of Community Importance (SCIs), countries whose wildlife credentials the average Englishwoman in the street might naturally question, e.g. France, Italy and Spain, have done much better (8.5%, 14.3% and 23.0% respectively) and it is little Slovenia, only a member of the EU club since 2004, which leads the way with a whopping 31.3%.

It's sometimes fashionable to think about the EU as a bunch of foreigners imposing their mad will on us sensible British, but this is far from the truth. The EU numbered only nine states when the Birds Directive was adopted – France, Germany, Italy, the Netherlands, Belgium, Luxembourg, Ireland, Denmark and the UK. By the time the Habitats Directive was adopted the EU had only grown to 12 members with the addition of Greece, Spain and Portugal. All of the other 15 Member States have had to sign up to two nature directives over which they had neither influence nor the opportunity to tweak to suit their particular circumstances. The Maltese and Cypriots had to sign up to bird protection laws that they would never have invented themselves and the Poles and Slovenians needed to set about designating areas of land under a Birds Directive that was invented a decade before the Berlin Wall started to fall.

Most of the Habitats Directive was actually written by a British civil servant who was then at a bit of a loose end, but who was a wildlife enthusiast. That civil servant was Stanley Johnson, the father of the current Mayor of London, Boris Johnson.

Stanley used to be a Eurocrat, and an MEP, but after standing down from his seat in the European Parliament he was rather adrift. The UK civil service didn't quite know what to do with him, so he wrote the Habitats Directive. This is, after all, the man who dashed off a love poem called *May Morning* one night, which then won the Newdigate Prize for poetry whilst he was an Oxford undergraduate. The Habitats Directive is certainly longer than the 300 line limit for the Newdigate but was a way for Stanley to express his love for nature. Incidentally, it was another couple of Brits, Alistair Gammell from the RSPB and Simon Lyster then with the WWF, who worked with Stanley to sell this draft Directive across the rest of the EU. So, it was actually a handful of Brits who wrote and promoted this piece of 'EU', and therefore 'foreign', environmental legislation that so annoys George Osborne right now.

These SPAs and SACs represent the very best wildlife sites across the EU. Many of the places where you have watched birds, or wanted to watch them, are in the list of designated sites – from the bustard-rich plains of Spain to the Danube Delta, and from the seabird colonies of Shetland to the saltpans and wetlands of the Camargue. The EU joins you in recognising these sites as part of our common European cultural inheritance.

This being the case, it is quite difficult to do things in Natura 2000 sites which will damage or destroy them – and quite rightly so in my opinion. Nobody would think they could pop into The Louvre and scribble on the Mona Lisa or knock over the stones at Stonehenge to build a factory, and so it should be with nature. There is no need for nature conservationists to be apologetic about the protection that has been given to the very best of the continent's special places for wildlife and there is an ongoing need to make sure that the protection given on paper is transformed into action and that what the bits of paper say is not ignored in the future.

If development is proposed in a Natura 2000 site then it is incumbent on the developer to show that there will be no impact on the nature conservation quality of that site (which is often the case) or to demonstrate that the development has an imperative to override public interest (and therefore should go ahead). Overriding public interest is quite a high test as it depends on showing that there is no available alternative – but since we are talking about the best wildlife sites on the continent, that doesn't seem unreasonable. If a development is of overriding public interest and goes ahead there is still a requirement to replace the wildlife interest of the site elsewhere, to maintain the wildlife stock of the EU, and that may add considerably to the development's cost. And remember, we are only talking about less than 10% of the UK land area being affected by these sites, and that they are of the very best judged in a European context.

Overall, given that appropriate management is likely to occur in SSSIs, and despite some worries about the adequacy of coverage in the uplands, and a dire lack of activity offshore, the protected area network in the UK would get pretty high marks from the bird population of this country if they had a voice on the subject.

Thames Estuary

Once or twice I've been in a plane flying over London and the Thames Estuary and looked down to see a large area of grassland below me. It is quite striking from the air as you spend time looking down on houses and roads and factories whilst circling and waiting for a landing slot. These are the North Kent Marshes and stretch from Cliffe in the west to Allhallows in the east. Looking down, I remember reading about them in my much-thumbed copy of John Gooders' *Where to Watch Birds* as a lad in Pensford, never having been to Kent in my life, and wondering whether I would ever visit these places with their wintering waders and waterfowl.

I already knew something of these marshes from Charles Dickens' 13th novel, *Great Expectations*, which I read as a child and had seen on TV, probably as one of those 'improving' classic serials that used to appear on a Sunday evening. Near the beginning of *Great Expectations* the young hero Pip is accosted in the churchyard, based on the one in Cooling village, by the very scary, escaped convict Magwitch (who even had a scary name). Magwitch had crossed the foggy marshes after escaping the prison ships – hulks in the Estuary, though we are not given his bird list. The encounter is on Christmas

Eve and the next day Pip earns Magwitch's eternal gratitude by returning to the churchyard and furnishing him with food and a file to free himself of his chains.

The marshes have as strongly defined a character as any in the book: 'the dark flat wilderness beyond the churchyard, intersected with dykes and mounds and gates, with scattered cattle feeding on it, was the marshes; and that the low leaden line beyond, was the river; and that the distant savage lair from which the wind was rushing was the sea.' The encounter with Magwitch stays with Pip throughout his life.

Thanks to the suggestion that the North Kent Marshes would be an excellent place to build London's fourth airport, there was a time when I was nipping down to Kent almost all of the time and this place will stay in my memory all through my life too. The government carried out a strategic review of airport capacity in southeast England and concluded that the North Kent Marshes – between the RSPB nature reserve at Cliffe in the west, through the High Halstow Marshes to the RSPB nature reserve of Northward Hill in the east – would not just be a good place to fly over, but also to land on a four-runway hub airport to rival Heathrow. Standing near the ancient heronry of Northward Hill and looking over the flat marshes away to the north and to the Thames, with built-up Essex on the other shore, it was difficult to imagine that the plan was to turn this semi-wilderness into noise and smell and concrete and metal.

It should not have escaped the government's notice that the North Kent Marshes are designated under domestic and European legislation. These marshes are designated SSSIs, part of an Environmentally Sensitive Area (ESA), a Ramsar site under the Convention for migratory waterfowl and, most importantly of all, an SPA under the Birds Directive. After an 18-month campaign waged by the residents of north Kent, the RSPB and other national wildlife NGOs, at local and national level, the government published an aviation white paper which did not support Cliffe as an option. Having taken the time to study the legislation the government acknowledged that:

> *...the potential benefits of developing a major new airport at Cliffe would need to be considered in the context of its significant impacts on important wildlife habitats. Moreover, the internationally important status of some of the habitats under European law mean that any potentially adverse effect would require the Government to demonstrate that it had considered all reasonable alternatives.*

In light of the consultation, the Government is satisfied that there would be reasonable alternatives to Cliffe…

The Government of the time learned that it was not possible simply to ignore environmental protection issues and the RSPB learned an awful lot about working with local communities on threats of this kind. The RSPB appointed the tireless Perry Haines as a local campaign co-ordinator and we immediately garnered lots of community support from local people whose lives would be destroyed by such an enormous development on their doorsteps – and in some cases, on top of their homes.

Some of the friendships made at that time have endured. Gill Moore and Joan Darwell, almost always seen together and so stamped on one's mind as 'JoanandGill', differ considerably in height, especially when Joan has her high heeled boots on, but are united in passion and dedication for the North Kent Marshes. I remember 'JoanandGill' telling me that they had not really known or appreciated the wildlife riches on their doorstep until they were potentially going to be lost. The herons from the large Northward Hill heronry were considered annoying visitors to garden ponds until their presence formed one small brick in a massive wall of wildlife protection for the area. And I remember many discussions with George Crozer in the Red Dog pub in High Halstow where we planned and plotted and sometimes despaired but eventually celebrated. The most moving aspect of the campaign was forging links with the people on the ground, those who would be most affected by any new airport.

I often thought, of course, of people in other parts of the country who might be affected if we won the battle not to build an airport here and it went elsewhere, but that was not something I could do much about. I was there to protect internationally important wildlife areas from an airport proposal that was wrong-headed and I was glad that that meant helping people like Gill, Joan and George along the way.

When in December 2003 it was clear that Cliffe was reprieved, the most moving event was attending a thanksgiving and carol service in a local church and realising that the community felt such huge gratitude to the RSPB for what we had done, and that that gratitude would form the basis for a better future for wildlife here for many years to come, provided that we built on the good will of the moment.

The real reasons for the ditching of Cliffe as an option were manifold. We mounted a media campaign that politicians could not ignore and backed it up with a petition of 160,000 signatures from all over the country asking the Transport Secretary, Alistair Darling, to abandon the Cliffe option. Local resident, Jools Holland supported the campaign. Actually, the aviation industry wasn't particularly keen on Cliffe as a site – it was an expensive and technically difficult project and the area would need a huge investment in rail and road infrastructure to function – it was the airport to nowhere in many people's minds.

Also, it didn't escape everyone's attention that a site recognised for the thousands and thousands of birds which visited it might not be the safest place to be landing aircraft full of people. Although bird-strike is a ridiculously overblown risk it does take a certain degree of recklessness to choose a bird-rich estuary as the favourite site for a new airport.

One morning I was making some of these points in a TV interview with Countryfile's John Craven in the sun on a small hillock at Northward Hill. It was a lovely day and the marshes spread out below us looked at their most inviting and least forbidding. Looking north across the marshes and the Thames, the industrial pipelines and flares of Canvey Island were shining in the sun. Above us a small aeroplane was circling and returned on a regular circuit of two minutes or so which was also the length of time that the interview should take. This meant that every time we repeated the interview it was interrupted by the return of the plane and the soundman would shake his head and say 'No, no good. You'll have to do it again'. For a few times this was just fun but as the number of re-takes passed a dozen I remember John Craven lifting his face to the sky, shaking his fist at the small plane and letting loose a stream of invective that, if broadcast, would have changed his image with the viewing public!

In autumn 2002, the RSPB held a Parliamentary reception for MPs and peers. Our speaker was the former Foreign Secretary, then Leader of the House of Commons, Robin Cook, who a few months later resigned from the Cabinet over the Iraq war. I spoke to Mr Cook about our worries over the government's plans and it was the Dickens connection that really brought the issue alive in his mind. I wonder whether he put in a word afterwards.

It would be nice to say that the No Airport At Cliffe campaign won a great victory, that local residents and national wildlife NGOs won the day and that everyone lived happily ever afterwards – and maybe we all will. But the price of wildlife riches is eternal vigilance and those areas of wild marshland, which look to me like a wildlife hot-spot, will always look to others like wasted areas ripe for economic development. The airport idea has not completely faded but now seems to be centred on the Thames Estuary itself – a proposal that will have most of the original problems of an airport in this area in spades – and there is always talk of a new Thames road crossing. There will always be someone wanting to make a fast buck at the expense of the wildlife that future generations could enjoy.

Lewis windfarm

AMEC and British Energy might feel that their application to cover an SPA on Lewis with hundreds of wind turbines was encouraged by government, and this is indeed a problem. The Department for Environment, Food and Rural Affairs (Defra) does not have great clout within government and other departments sometimes appear to think that their interests can lord it over the environment. The need for renewable energy has sometimes, it seems to me, encouraged other government departments to encourage the windfarm industry to press ahead with doomed projects.

If internationally important peatland sites were threatened by a project for a new supermarket, then the scheme would have stood no chance of approval. Because it was for a windfarm, some decision-makers saw it as a green option, despite its impact. And one of those impacts – rather ironically – would be the release of huge amounts of carbon from the damaged peatbogs themselves.

We met for a high-level chat about this windfarm proposal with our friends and colleagues in Greenpeace, whose work was much more focussed on climate issues and far less on wildlife issues than our own. Over a pizza in a north London restaurant, we tried to find a way not to fall out over this issue and we did, but only after some fairly harsh words were exchanged on both sides. We subsequently kissed and made up!

Even in its modified and revised form the Lewis windfarm proposal was for 181 wind turbines to be built across the Lewis Peatlands SPA and affected the Lewis Peatlands SAC as well. The area is, in some ways, the western extension of the Flow Country with somewhat similar birds: black-throated diver, red-throated diver, greenshank, dunlin, golden plover, golden eagle and merlin. Only if this proposal had no impact at all on the internationally important wildlife of that site would it make any sense at all as a serious planning application. And it was inconceivable that building access roads across the deep peat, sinking the foundations for 181 wind turbines, and the increased disturbance through traffic and maintenance visits would leave this fragile wetland and its wildlife unaffected. If the developers had a single ecologist on their staff then they ought to have known all this and thought very hard before risking their money and their shareholders' money on tilting at the windmills of a Natura 2000 site.

I don't have much sympathy for any chief executive or politician who doesn't bother to know their job well enough to understand the ultimate or reasonable constraints that long-existing legislation puts on their plans. If you can't be bothered to know your job properly, then you are likely to come a cropper – and it isn't the job of wildlife NGOs to spare you any of the pain that will surely follow your lack of professionalism.

Severn Estuary

Brean Down was a place I used to visit regularly with the Bristol Grammar School Field Club, a promontory that extends from the Mendip Hills into the Bristol Channel towards the coast of south Wales. We would visit early on a late April morning to look and listen for arriving migrants. I saw and heard my first grasshopper warbler on one such trip. The site is now managed by the National Trust but I learned much later that it had once been an RSPB nature reserve.

At the landward end there is a cliff on which we used to see a pair of nesting ravens, and at the seaward end we would stare out to sea and see very little. Once, however, two birds rounded the point, close in, flying up the estuary and we, a group of teenage boy birders, were all thrilled to see a Manx shearwater so close and so well. This was a good bird to see and it made the day. After the shearwater and its brown companion bird had gone we realised that half the party was convinced that the second bird had been a female mallard while the other half believed it had been a curlew. This incident sticks in the mind 40 years later as an example of the type of trick that the mind can play. If the brown bird had been the only bird flying past there is no doubt that we would all have known what it was, but in the presence of a star-bird, the shearwater, which had almost our full attention, the second bird had been terribly misidentified by half of us – but which half?

Staring out from Brean Down one can see two islands, Steep Holm and Flat

Holm. Steep Holm is part of Somerset and Flat Holm is in Wales. Neither is very easy to visit and both have lots of gulls nesting on them but very little else. Both have always seemed rather mysterious to me and yet if the plans for a Severn Barrage had gone ahead there would have been a 12-mile wall of concrete stretching from the point where I saw a Manx shearwater fly past to the island of Steep Holm before making land at Lavernock Point in Glamorgan.

The Severn is an interesting estuary with the second-largest tidal range of any in the world. Huge quantities of water and sediment flow past Brean Down every day thanks to the pull of the moon on the Earth's seas and oceans. Ever since Victorian days there have been ideas to harness this energy, which seems endless and free. In a world where greenhouse gas emissions need to be cut dramatically, then it came as no surprise that the Severn Barrage came back onto the agenda.

The possibility of building such a barrage was announced at the Labour Party Conference in September 2007 by the Secretary for Trade and Industry, John Hutton. You should always doubt the motives of government announcements timed to coincide with their particular party conference as grand-standing of the worst order, and it had been rumoured beforehand that Gordon Brown, the Prime Minister, had considered making the announcement himself. This followed a report earlier in the year by the Sustainable Development Commission which had highlighted the energy-producing possibilities of a barrage, but had also high-lighted that any such project would need to comply with the Birds and Habitats Directives 'without reform or derogation' – in other words, 'no bending the rules!'.

The Severn Estuary is an SPA because of its wintering waterfowl, such as the Bewick's swans and white-fronted geese that I used to go and look for at the WWT reserve at Slimbridge, but it is also an SAC owing to its importance for migratory fish such as eels, twaite shad, lamprey and salmon. The Severn, Usk and Wye are salmon rivers of renown, although much diminished in richness now. I remember my father being given a Wye-caught salmon and bringing it home when I was very young – I got to stay up late as my parents dealt with the fish in the kitchen – and this was long before we had a freezer so it was fresh salmon on the menu for us, friends and relatives for a few days.

The RSPB never opposed the use of the Severn estuary for the generation of tidal power – it seemed like a good idea to us. What we promoted were the various alternatives to a Victorian barrage wall – tidal streams, tidal lagoons and smaller barrages in a variety of different places along the estuary. We also looked carefully at the case for the big barrage in terms of economics and potential greenhouse gas emissions.

But, if you put your head above the parapet then you might get shot at. The RSPB was only one of many environmental organisations concerned about the plans for a Severn Barrage but we were probably more feared than others because of our large membership and technical ability. The RSPB commissioned several important studies of the Severn which looked at the economics of the various options, the different locations for a barrage or barrages, non-barrage options, etc. as well as mentioning the 69,000 waterfowl that use the Severn Estuary and its particular importance as a refuge in cold winters for migratory waterfowl fleeing westwards across a frozen Europe.

But our measured and detailed approach was not always matched by politicians. The Energy Minister, Malcolm Wicks, unleashed an attack on the RSPB in a Welsh Grand Committee meeting at Westminster where he told MPs the RSPB was:

> *...clearly not understanding that unless we are prepared to take some courageous action on climate change the devastation of species will be truly enormous. It is the duty of a sensible NGO (non-governmental organisation) supported by the public that occasionally they say yes to projects and (are) not always seeking the comfort zone of saying no to a barrage, no to a windfarm, no to this, no to that...*

and:

> *There needs to be a responsibility and a seriousness in all organisations, especially the environmental ones.*

We replied as follows:

> *Check our website to judge how seriously we take climate change. Mr Wicks is ignoring the difficult issue facing the government over the cost of the Severn Barrage - the energy it produces can be produced at half the cost by other renewable technologies. Why should we*

> *spend £15bn, at least, on a barrage when the same amount of renewable energy could be produced at half the cost? Does Mr Wicks think that wasting £7.5bn is good government policy?'*

We also upset Mr Wicks by writing to his constituents who were RSPB members to point out to them what their MP had said as a Minister and to correct any misunderstandings of our position.

We brought some people over from the Netherlands to tell our decision makers about the Dutch experience in building a storm surge barrier in the Oosterschelde estuary in the late 1980s. The barrier was built to defend against flooding but its building resulted in losses of tourist income, damage to shellfisheries, loss of wildlife, and much larger than expected flood defence bills as a result of altering the ebb and flow of the tides in the estuary.

When, in October 2010, the Energy Secretary, Chris Huhne, announced that the barrage scheme would not go ahead the reason given was money and there was no acknowledgment of the ecological damage that such a scheme would cause. There were always more people in the queue of those keen to be paid to build a barrage, than those in the queue with £10–20bn of ready cash wanting to pay for such a scheme.

The Severn Barrage will undoubtedly come back on the agenda some time – there is an awful lot of renewable energy out there and there is a lot of money for someone to make too. I would like to see us tapping into all that energy in a modern way. I'm sure that there are future battles to be fought to ensure that nature isn't sacrificed for money and on the altar of a headline-grabbing announcement for a future politician.

Natura 2000 sites and their protection

There are a variety of myths about the Natura 2000 designations. These include the view that these Directives are all-powerful, that they stifle economic development, that they are implemented more strictly in the UK than elsewhere, and that they favour wildlife over people. None of these views rings true to me, but they are all promulgated by vested interests in industry and are sometimes accepted unquestioningly by politicians.

The Nature Directives do not prevent all developments, only ones which will damage the wildlife interest for which the sites were originally designated. Many developments do go ahead in Natura 2000 sites – some of which probably should have been denied in my opinion. The Cardiff Bay Barrage is an example of a flawed and unnecessary development which, in my opinion, destroyed more environmental value than it created economic value.

Many of the best outcomes have arisen when developers worked with wildlife NGOs to maximise benefits for both nature and economic development. Examples include an agreement in the Thames Estuary for a windfarm which would have affected wintering red-throated divers – by working with the RSPB the developers found that a smaller and staged scheme could go ahead, whereas their original proposal would have met fierce resistance. Such compromises are worth seeking, but they don't always exist and industry does not always look for them.

I wonder whether AMEC and British Energy would agree with my assessment, in retrospect, that the Lewis windfarm proposal was a very bad business decision? Why did they apparently ignore the lines on the maps which showed that their preferred site had the strongest European wildlife designation possible? And why did they not realise that pressing ahead would waste their time and money and could only succeed if the statutory system were utterly incompetent? Were they hoping that a renewable energy project would be set a lower threshold of acceptability than any other development – and if so how on earth did they come to that view?

It's not as though there weren't already existing case studies which should have warned them to back off. One such was Associated British Ports' application to build port facilities at Dibden Bay in Hampshire. ABP's was just one of several applications for a new port in southeast England at the time and the Secretary of State had to make some decisions in the absence of any strategic policy on port capacity. The danger was that applications to build new port capacity would have been agreed that would have far-exceeded the nation's needs. When ABP's plans were turned down the company's share value dropped by, as I remember it, £40m on the day, and took a while to recover. A more thorough understanding of environmental legislation might have spared the company and its share-holders this loss but it has struck me time after time that industry does not treat environmental legislation as seriously as it has learned to treat health and safety or employment legislation. Since then, and with different senior management, ABP has been one of the best port companies in its appreciation of the need to understand environmental legislation as part of its business.

Sometimes the Directives are portrayed as being pro-wildlife and anti-people. I've never seen this as being true. Let us imagine that the go-ahead had been given to build an airport at Cliffe. There is a big chance that by now, in the current economic climate, the money would have run out or never been found. Given the lack of enthusiasm for the project from the aviation industry it is possible that it would have been an economic disaster which would have benefited no-one. Add the risk of bird strike at the site and you could imagine that the Directives have prevented economic and human disaster as well as protecting an internationally important wildlife site.

And the Severn Barrage, if built, would have destroyed the salmon fishing on three rivers, finished off the elver fishery, greatly reduced the viability of the port of Bristol, disrupted the lives of many residents of north Somerset and south Glamorgan, and potentially increased flood risks in some coastal areas – and reduced the wildlife of an internationally important estuary. There are always costs and benefits on both sides and wildlife rarely stands alone on either side of the equation. Developers usually talk up their side outrageously in order to gain political support – they promise new jobs and economic development from projects that almost always under-perform in benefits and take longer and more money to build than was originally claimed. But it's certainly not wildlife on one side and people on the other – it's usually some people on one side and some other people and wildlife on the other.

At the time of writing, the Government has decided to review the implementation of the Nature Directives as the Chancellor, George Osborne, announced in his Autumn Statement in November 2011 that he was concerned that gold-plating of the Directives placed 'ridiculous costs' on British businesses. It is not clear what George Osborne has in mind, but it may (or may not) be projects like an airport over Dickens' coastal marshes of north Kent, the Severn Barrage, Lewis windfarm and Dibden Bay that he thinks should have gone ahead.

Lessons learned

A network of protected areas is bound to form a part of any nature conservation strategy. It is a sensible response to the basic requirements of animal and plant populations – which are not uniformly distributed over the Earth's surface. And once you realise that some places are much more important for nature than others then finding a way of identifying and protecting the best ones is clearly needed. Nature is reasonably well-served by the current system – and so are we as people.

But there is always money to be made from pouring concrete and the areas that are protected now will always look to some like perfect places for the next round of concrete-pouring.

So the conservation movement needs to be prepared to re-fight the battles again and again and again. And those battles will require an understanding of the economics and technical details of the developers' cases as well as, indeed more than, knowledge of the wildlife impacts of the schemes. For the best protected sites it will be the arguments about damage to the site and whether there are viable alternatives which are crucial in determining overwhelming public interest.

But for many of our so-called protected areas, the SSSIs, their value will depend on continued favourable management. Much of the money for that management comes from the wildlife NGOs who own them and from existing statutory agri-environment schemes. In straitened economic times both sources of money are under threat – or at least squeezed. Maintaining proper management of SSSIs is essential to maintaining wildlife richness on that small area of land so designated – and I predict that there are tough times ahead.

> *We do not inherit the earth from our ancestors; we borrow it from our children.*
> **Native American Proverb**

Chapter 7

Hope for farmland birds

Hail to thee, blithe spirit!
Bird thou never wert
That from heaven or near it
Pourest thy full heart
In profuse strains of unpremeditated art.
Percy Bysshe Shelley

CBC at Cambridge

I remember from my Cambridge days a CBC survey plot behind the rugby ground on Grange Road which stretched west into the Cambridgeshire countryside. Surveying this site was a task shared between the undergraduates and that meant that occasionally I would be out there with my pencil, map and binoculars at some ridiculous time in the morning after, no doubt, a late, student-like night. I didn't do a lot of birdwatching while at Cambridge and so this early morning task was a way of maintaining my link with birds.

The survey plot edged up to the back gardens of some houses and then stretched out across the arable fields so typical of East Anglian farmland. Not a stunning place for birds but that was the point – we were recording what was happening to the average bird in the average place. The best bird I recorded on this site was a quail which was singing repeatedly from the wheat one June morning, but my abiding memory is of the number of skylarks. To record a skylark on a CBC map you write a capital S with a dot thus; 'S.'. If you can see several skylarks at the same time then you can join them up with lines. And if you see them having a bit of a fight then you can surround them with little explosion lines. My recollection is that my map was quickly covered with 'S.'s and various types of line. If I picture myself recording birds on this site then I picture myself looking bemused and covering the map with skylark symbols.

At the time, recording skylarks was a slightly irritating chore but now the skylark has become a symbol of what we have done to our countryside. I imagine, and I will do this one day, that if I were to return to this site then I would be less perplexed by the abundance of skylarks because they would be many fewer in number. Overall, skylark numbers have fallen by over 50% since those times – fewer lines, less pencil to be used but less joyous song in the countryside.

Farmland bird baselines

But a decade passed before we really began to realise the scale of the decline in numbers of farmland birds. The BTO's book on bird population trends, published in 1990, was a breakthrough publication.

We began to realise that in the UK, particularly in southern England, farmland birds were bleeding from the landscape. Once abundant species, such as grey partridge and lapwing, were becoming unfamiliar birds in many parts of their former range. Tree sparrows and corn buntings were now much rarer. In a period of perhaps just 25 years we had driven millions of birds from the countryside and out of our lives. The scale of this loss is enormous and the CBC results show that it was concentrated in the period 1970–1980 although it is quite possible, probable I would say, that this decline was part of a much longer and even more dramatic decline that had been going on for decades.

One of the species that has suffered the most dramatic decline is the grey partridge and the person who highlighted its decline, and that of many other farmland birds, was the former director of the Game Conservancy Trust, Dr Dick Potts.

The first time I met Dick I remember him saying that he'd heard of me and had wanted to meet me. I felt just the same way about him, and with far better reasons. Dick's passion is grey partridges. He has studied them for decades and his work has been recognised as of the highest standard with a whole range of honours and prizes. And he is a very nice guy. I remember spending a day with Dick on his Sussex study area and being just delighted by how much he knew about the way that farmland ticked and how wildlife fitted in or was being squeezed out of the picture.

Because the grey partridge is a quarry species we have bag records of shot birds going back long before birders had got their act together enough to count them. There are quite a few differences between counts of dead birds and counts of living ones, but in the absence of the latter the former tell one quite a lot – although there is always a nagging doubt about whether hunting effort (e.g. number of hunters, hours shooting) or efficiency (e.g. better guns or ammunition) change through time. But looking at the bags of certain estates the grey partridge must have been phenomenally abundant.

The Game and Wildlife Conservation Trust (GWCT) have compiled bag records of grey partridges that go back to 1804 – if only we had a BBS covering such a long period. If we compare the current situation with that of 200 years ago, only about one fifth of the number of grey partridges are shot in the countryside today. That probably means that there were at least five times as many live partridges two centuries ago. But it would be wrong to think that there has been a long, slow decline – in fact the fortunes of the grey partridge soared in the 19th century as the enclosure of land provided hedgerow nesting habitat literally alongside an increased acreage of arable land. And so, at the beginning of the 20th century the number of grey partridges which were shot was about three times the number at the start of the 19th century and about ten times that shot nowadays. At around 1900, two million grey partridges were being shot in the UK each year. Since the GWCT was founded over 75 years ago partridge numbers have declined despite the excellence of the research that has been done by that organisation to try to provide a better future for the bird chosen for their logo.

The lapwing is another species whose numbers have tumbled. Plovers' eggs make a tasty dish and I remember being out birdwatching one winter's day in the Nene Valley, near where I lived, and coming across an older man, who was looking at a circling flock of lapwings. It was a crisp, sunny day and we watched the lapwings together with pleasure and then started talking about them. We realised that we lived near each other but my companion, unlike me, had been born and bred in this part of the world. He recounted that in his youth he and his brother used to borrow a baker's tray in spring and bring it down to the fields overlooking the valley where we stood to spend an afternoon looking for lapwing eggs. We must have been talking about the late 1940s, or a little later, and he said that they could easily fill the tray with eggs. I thought of my own children and couldn't imagine that, even with a fast car, they could find that many lapwing eggs in the whole county these days. Such is the scale of loss of this once common farmland species, that it is now rare indeed in many parts of England, Wales and Northern Ireland – and getting rarer even in northern England and Scotland.

But clearly some farmland birds have increased in numbers – the wood pigeon and jackdaw, for example, are now much more common than they used to be. So in a situation where some species have increased and some have decreased, how does one interpret the changes overall and how does one bring this message home to the public, farmers and decision makers? What is the right way to look at change in farmland birdlife?

Back in the 1990s the RSPB and the BTO worked together to produce an index of farmland birds that would capture the essence of changes in species populations. The Farmland Bird Index (FBI) was born. It takes those species recognised as farmland species in the *Atlas of Breeding Birds* (and, importantly, those whose numbers are monitored annually) and expresses their combined population changes in a single number. The 19 species which contribute to the index are; lapwing, wood pigeon, stock dove, turtle dove, rook, jackdaw, starling, skylark, kestrel, grey partridge, whitethroat, yellow wagtail, linnet, goldfinch, greenfinch, tree sparrow, reed bunting, yellowhammer and corn bunting. The latest figures for the FBI, published in November 2011, show that for the UK as a whole, the FBI reached its lowest ever point in 2010 (the last year for which the data have been collated and analysed). Farmland bird populations had fallen to half the levels recorded in 1970, according to the official government statistics. Of course, if you took out the species which had decreased the most (corn bunting, turtle dove and grey partridge, which were all at less than one tenth of their 1970 levels – 90% declines!) things don't look so bad, but if you take out the increasing species such as wood pigeon and goldfinch, then things look even worse. And that's what the index is for – to take a suite of widespread species dependent on farmland (mostly lowland farmland species because those are the ones for which annual data are most easily available) and monitor their overall changes in abundance.

Losing half of the birds inhabiting farmland in a generation is a dramatic impact and one which hits at the heart of whether current agricultural practice is sustainable. What type of farming system do we have that drives wildlife from the countryside?

Why?

The fact that many farmland birds declined in numbers at the same time gave a substantial clue as to where to look for the causes of those declines. The species in decline were a mixture of summer migrants and residents but the fact that most were residents suggested strongly that at least some of the causes were based here in the UK and that we couldn't blame French hunters or land use change in the Sahel for all of what we were seeing here at home. The declining species were widespread across the UK and so it didn't seem to be a local problem particular, say, to East Anglia or Lancashire. These population declines were happening everywhere. And although we tended to emphasise the declines, some farmland species, such as wood pigeon and jackdaw, showed marked increases in numbers and so were presumably benefitting in some way from the factors that were harming the majority of species. Clearly, we should look to farming for the reasons behind these species declines.

Farming practices have changed enormously over time, and the 1970s and 1980s were decades of particularly stark change.

Thanks to the emergence of new herbicides it became possible to sow cereal crops in the autumn, soon after the previous crop was harvested, instead of waiting until after Christmas for a period of suitable weather to get onto the land. For the farmer this made life simpler and reduced the uncertainties of sowing but it also meant that the crop had a head start whenever the spring weather arrived and could grow more quickly and provide higher yields. For wildlife it meant that stubble fields largely disappeared from much of the countryside – and they had acted like massive winter bird tables for skylarks, buntings and finches who could feed on weeds and grain spilled from the previous harvest. And when the spring came ground-nesting birds found it more difficult to nest in arable crops because they were thicker and taller than they had previously been. No-one changed agricultural practice deliberately to disadvantage farmland birds. Things would have been very different if these changes had benefitted farmland birds, but in the event they were generally bad for them.

A guaranteed price for your produce encourages specialisation and through the individual choices of thousands of farmers the countryside became more polarised. In the green west livestock farming was the norm, while in the yellow east wheat, barley and oilseed rape became the crops of choice. Most birds have varied diets – almost all need insects for their young in the breeding season but feed much more on seeds in the winter. Mixed farming systems are ideal for many birds, enabling them to meet their needs throughout the year. A mixed farm will probably have decent hedges because they are of value in preventing your stock from straying. Animal dung encourages insects and out-wintered stock are fed with hay and have straw bedding, both of which will contain some seeds and other edible matter. Spring sown cereals allow lapwings to nest, whose chicks can move to nearby grassland to feed after hatching. The occasional stubble field may provide food right through the winter. This mixture of opportunities is much reduced on a silage-dominated dairy farm and on a pure arable farm with large fields. Again, what perfectly suited thousands of farmers did not suit millions of birds.

Two main trends manifested themselves in the countryside – specialisation and intensification. Both were indirectly encouraged by the subsidy system of the Common Agricultural Policy – both were the product of perfectly rational decisions by individual farmers – and both were bad for wildlife on farms.

But in order to get decision-makers to take these declines seriously we needed more evidence of their causes, and some solutions to suggest. A large number of studies of individual species were carried out, largely by the RSPB, during the 1990s in order to better understand the situation and to provide evidence for the causes of the declines – and, just as importantly, to provide workable solutions for modern farmers. I'll mention several of them here, but touch in more detail on just three. Two were carried out by the RSPB but for the first let us go back to the work of Dick Potts and the Game Conservancy Trust (GCT) on grey partridges.

Grey partridge

Dick Potts and GCT had started work on farmland birds, or at least that one farmland bird that was formerly the most important quarry species for shoots, long before the rest of us got into the act. The study is a classic and combines field research, experimental manipulation, mathematical modelling and a big pinch of inspired intuition.

The key factor in maintaining grey partridge numbers is increasing chick survival –and the young chicks eat insects such as sawfly larvae, crickets and plant bugs. If there aren't enough insects then the chicks starve and you won't have any grey partridges. The decline in chick survival through the 1950s and 1960s was clearly caused by the big fall in insect food available in arable fields. These declines in insect numbers were probably mainly caused by the increasing use of herbicides which eliminated plants (weeds to the farmer) that provided food for the insects on which the grey partridge chicks depended.

Herbicides are clearly important for maintaining high crop yields and it's no good to ask farmers to stop using them – except if they are going to become wholly organic which is still really a minority practice. But Mike Rands (who later became the Chief Executive of BirdLife International) and

Nick Sotherton (still of the GCT) carried out some neat experiments showing that leaving field margins unsprayed – less than 2% of the field area – could dramatically increase grey partridge chick survival and lead to increases in grey partridge populations. They also showed that there was no noticeable reduction in overall crop yield as the area affected was small and tended to be the least productive part of the field.

This treatment – the Conservation Headland – was promoted by the GCT to their shooting members, and to farmers generally, for many years, alongside beetle banks in the middle of fields and a range of other grey partridge-friendly measures. But the decline in grey partridge numbers has continued. I'm afraid it's an example of great research not being sufficiently widely applied on the ground – and not the only such example. This was, in today's parlance, a Big Society solution which depended on people doing the 'right thing' even though they don't actually have to. It hasn't worked for the grey partridge – at least not yet.

And, slightly embarrassingly for the GCT, they started with a very small population of grey partridges on their demonstration farm in Leicestershire and unfortunately lost them. So far this population has not recovered despite all their efforts, including a lot of predator control measures.

Skylark

Shelley was a radical as well as a romantic poet and I am pleased that it was he who wrote 'Hail to thee blithe spirit'. But skylarks have had nothing to be happy about in recent years. This is, in some ways, the archetypal farmland bird. It occurs in arable and pastoral landscapes, is common and numerous, and feeds on insects in summer and seeds and vegetation for much of the rest of the year. It even nests on the ground in the middle of crops. There could be no better indicator of the health of the relationship between wildlife and modern farming than the skylark – and its numbers have halved over recent decades.

I sometimes use the skylark story as a classic example of how to 'almost' conserve a widespread species for which nature reserves and protected areas cannot work. And the 'almost' is important here.

The decline in skylark numbers came as a surprise to many nature conservationists and is still not really accepted by some farmers, who will point out that there are still skylarks on their farms and say that nothing has changed. Well, actually, it is clear that a lot has changed for the skylark.

Paul Donald, who has written a book on skylarks, was employed by the RSPB first of all to analyse BTO data on them, and then to carry out fieldwork to try to find solutions to the species' plight. He showed, very convincingly, that skylarks did better in areas of spring sowing than the autumn sowing that had taken over in the 1970s and 1980s. But his research also explained those differences. Where crops were sown in the autumn, skylarks were excluded after their first broods as the crop became too thick for them to nest in. Instead they were pushed to the tramlines (the routes down which agricultural machinery travels through the crop) where they were vulnerable to predators such as foxes strolling through the field using the tramlines as an easy route, or simply to the return of tractors on their visits to spray with chemicals.

There were other factors at play as well, such as the loss of stubble fields and the use of pesticides (both linked to the switch to autumn sowing). The ideal solution for skylarks would be a return to the agricultural practices of 30–40 years ago. This was clearly not going to happen and so we looked for a 'quick-fix'.

The fix that was tried was to establish skylark 'scrapes' or 'patches' – small unsown areas in the middle of the crop field. These patches always look rather large as you look out across a field, but in fact they occupy a tiny area of land so their impact on crop yield is negligible. We thought that leaving bare patches would provide skylarks with spots in which to nest, and sometimes they do. But more often they feed in them (as do other farmland birds) and also use them as landing pads from which to enter the rest of the crop.

The idea, which was trialled at a number of farms across the country including the RSPB's own arable farm, was highly successful. Fields with skylark patches had more skylarks for longer through the breeding season and they produced more young than those fields without them. Skylark patches worked!

And so the 'almost' comes from the fact that the farming community has not adopted skylark patches as a routine part of farming practice as we hoped they would. This is despite skylark patches being a part of the English agri-environment package – you will get paid for leaving these small patches in which wildlife can thrive. But not enough, not nearly enough, farmers have adopted them to make an impact. There has been a distinct lack of action from those leading the farming community, such as those at the top of the National Farmers Union, to adopt and promote this proven and effective remedy for the skylark's problems despite the research having been done, the effectiveness proved and the money being there. It's difficult to see how conservationists could do more to provide an easy way for farming to look good than this, and yet the majority of the arable farming community has shunned the option and shunned the skylark. But nature conservationists are still standing and so is the skylark, so let's just assume that success is postponed rather than lost.

Cirl bunting

Cirl buntings are familiar birds in the olive groves and orchards of southern Europe but only just creep into southern England, into south Devon in particular. But back in the 1950s this bird was much more widespread and occurred as far north as Herefordshire. Back then one might even have glimpsed it on a train journey into London from the RSPB's Headquarters at Sandy. In the 1970s there were still a few pairs in Somerset. Although I never saw them, I remember boyhood records from Worle golf course, Crook Peak and Cheddar Reservoir.

The cirl bunting was an early indicator of what would happen to many other farmland birds in later decades, but when Ken Smith and Andy Evans started working on the species it seemed like the study of a rare species rather than a precursor to many other studies of farmland birds.

Despite its previously wide range the cirl bunting became localised to the hills and valleys of south Devon – why there? Observations of the birds in winter showed that they were concentrated in the few stubble fields which still existed in this farming area. A very small number of stubble fields were feeding a large proportion of the UK's cirl buntings and so negotiations began with individual farmers to maintain this farming system in the key localities whilst more information was gathered.

In summer, cirl buntings nest in thick hedges, such as still remained in south Devon but had disappeared elsewhere, and feed their young on large insects – particularly grasshoppers. Fairly unusually for birds, it is the late cirl bunting nests that are the most productive – those built in July and August, when juicy grasshoppers are most abundant.

The skill and ingenuity of the researchers who unpicked the details of this story were amazing but in retrospect, as is often the case with good science, the results look pretty obvious (they never do when you are starting the work!). Cirl buntings need seeds in winter, safe nesting sites and lots of grasshoppers or other large insects in summer. Variations on these three needs – the 'Big Three' – have become the standard mantra both for farmland advisors talking to farmers and for policy advocates advising civil servants and politicians.

So now we understood why cirl buntings were hanging on in south Devon and had been lost from many of their former haunts. In south Devon the combination of soil type, climate, and steep wooded valleys, and probably the ethos of an attractive landscape and its benefits for the tourist industry, combine with an inherently conservative farming community to make changes in farming practices less dramatic and less rapid than elsewhere in the country. The farming system has inhabited a sort of a time warp, allowing cirl buntings to survive.

But the march of change continues and tends always to change in ways which would disadvantage farmland birds, including cirl buntings. The RSPB worked with farmers and decision-makers to ensure that the grants available for wildlife friendly farming favoured cirl bunting-friendly options wherever possible – and this wasn't straight forward as there were times when it looked as though the incentives might well swing farming against the cirl bunting in its last refuges.

The introduction of compulsory set-aside helped the cirl bunting and with farmer goodwill, set-aside, and well-targeted agri-environment funding the cirl bunting population grew from 189 pairs in 1989 to 862 pairs in 2009. That's a marvellous increase considering that the reasons for the decline were not well understood at the outset and that solutions depended on a change to government funding. The cirl bunting story is one of success and I hope that future changes to farm payments don't compromise continued success. But for now we must take heart that the cirl bunting story shows what can be done by farmers, scientists and nature conservationists working together.

The cirl bunting story gives hope that similar successes can be achieved for other, more widespread, farmland birds before their populations sink to similarly low levels.

These three examples, the grey partridge, the skylark and the cirl bunting are just three from a whole range of declining farmland birds the study of which has led to a deep understanding of the problems they face. Each has benefitted from completely different practical solutions – conservation headlands for partridges, crop patches for skylarks and mixed farming for cirl buntings. We also know that lapwings declined to a large extent because of the loss of mixed farming and spring-sown cereals; that turtle doves have lost the arable weeds on which they depend; that starlings have declined in the east because of the loss of grassland; and that corn buntings like barley and suffer from their later nesting being curtailed by earlier harvests. In the west of the country the move towards cutting for silage has led to increased fertiliser use and earlier and more frequent mowing which reduces invertebrate densities and destroys more nests. We know lots about these declines and the causes can all be bundled under the rather inelegant term, agricultural intensification.

Intensive agriculture is not a popular term with the farming community. It is short hand and therefore doesn't capture all the things that are going on. Industrial farming might be a better term but that isn't likely to be popular either. I won't call our current farming system 'modern farming' because I am sure there is a better and more modern sustainable farming system within our grasp. But time after time when people have looked for the reasons for declines of farmland birds they have found that changes in agricultural practices are the root cause of the changes in bird numbers.

Not just us and not just birds

The loss of birds from the countryside is a very personal issue to me. It affects the quality of my life as I spend a lot of time walking in the countryside and it's one of the things about England that I love. Every time I go for a walk my life is diminished by the fact that I am unlikely to see a single tree sparrow or corn bunting, as well as seeing many fewer of a great many other species.

But my personal feeling of loss is shared by those who walk in their own countryside whether in the plains of La Mancha, in a Bavarian meadow, or in an olive grove in Greece. It's not just the Euro, the banks and the economy that are in trouble across the continent, it is the natural capital of farmland birds. In every European country farmland birds have declined during our lifetimes. A study by Paul Donald and others (yes, he of skylark fame – clever guy, Paul) looked at European farmland bird trends and essentially produced a Farmland Bird Index for each country. They found that the index had fallen everywhere – from the Straits of Gibraltar to the Bosphorus, and from the Baltic to the Irish Sea. Europe is united in losing its farmland birds.

But the study highlighted two things of interest – one particular and one general. The particular was that the UK led the way in having got rid of its countryside birds – we had a higher estimated FBI decline than other European countries. Given the accuracy of the figures we can still be sure that we are in the very top league when it comes to losing our farmland birds –

and that is a sobering thought and one to keep in mind whenever farmers' leaders talk about environmental matters. But the general point is much more interesting and important because it explains why the UK is at the top of the league of farmland bird loss and why my walks in the countryside are so diminished in natural pleasure. Paul and his co-authors showed that the size of loss of farmland bird populations for each European country was related to the productivity of its agriculture – the countries which produced most wheat per hectare or most milk per cow were the ones which had lost most larks, buntings and shrikes.

This finding both answers and poses a question at the same time. It answers (not on its own, but taken with lots of other evidence) the question about why farmland birds are declining, not through studies of individual species but through the broad sweep of continent-wide trends. Productive agriculture, in the way that we practise it in Europe, is inimical to wildlife. And that poses an uncomfortable question for those of us who love nature – do we have to choose between a countryside that is productive for food and one which is productive for wildlife? I'll leave that question hanging for now, but we will come back to it (and don't become too depressed just yet – there's a happy ending).

They say that travel broadens the mind, but I think that sometimes it can narrow it. I remember, as a student, once travelling through France and Spain on a train in the company of an English girl who spent the whole time pointing out things that she didn't like as she looked out of the window. At Victoria she had seemed a delightful chance encounter who might make the journey ahead more enjoyable, but by Lisbon she was an annoying xenophobe who had marred the pleasure of new sights and sounds (although the smells of Spanish trains then weren't that enticing). I was trying to spot a hoopoe or a Montagu's harrier and I did see some new birds from the train – it never ceases to amaze me how rich in wildlife much of the farmland in continental Europe is compared with our own. Stand in a field in another European country and you will usually find nature on hand in a way that England, or much of the rest of the UK, cannot match.

Go to Spain and stop by the side of the road in an unpromising looking piece of farmland and you will hear corn buntings singing with an intensity rarely heard in the UK. Or go to eastern Europe and see red-backed shrikes in the fields as part of the natural scheme of things, eating grasshoppers and other large insects just as they were in the fields of southeast England a century ago.

If you do go to these places and are a better all round naturalist than I (not difficult) then look at the invertebrate life buzzing around you, and the flowers beneath your feet. You'll find that the birds are only the most obvious manifestation of a much richer fauna and flora than you're used to seeing when walking in your own countryside back home.

I wish we had the data from the UK and across Europe to produce the Farmland Insect Index, Farmland Flower Index, and even the Farmland Earthworm Index or Farmland Soil Carbon Index. I wonder what they would tell us about our impact on those aspects of the ecology of our food production systems. My strong feeling is that they would tell an even more depressing story than do the bird data – birds are pretty tough and adaptable. Maybe the wider data are out there and maybe someone should have a look at them.

But we do know something about the losses of non-avian life from farmland. There's one interesting case which tells us quite a lot about the fragility of nature. How often do you see a cornflower, a corncockle, a corn parsley, a corn buttercup, a corn spurrey or a corn marigold on your walks? These are all, now, rare on arable land. They used to be much more common and their names, along with pheasant's-eye, field gromwell, field pansy, shepherd's needle and field madder, illustrate the fact that they were once much more commonly associated with arable farming. In fact, there is a large group of threatened plants which are sometimes called rare arable weeds – because they used to be so abundant that they were a problem for farmers but are now so rare that they are a worry for nature conservationists.

The reasons why these once-common species are now so rare overlap considerably with the reasons why birds have declined so much – increased use of herbicides and nitrogen, changes in cropping dates, and so on. Modern herbicides have become so efficient that very few weed species can survive in what are effectively sterilised crop fields. A few, such as wild oats and cleavers, are still big problems for the arable farmer and one known as black grass by farmers and rather more attractively as slender foxtail by botanists is now one of the biggest problem plants for the arable industry.

The consequences of intensive farming are an issue mostly pursued by bird conservationists over the last couple of decades. Birds are relatively tough and disappear slowly enough that observable losses are still noticeable and alarming. But it seems to me that in fighting for a more sustainable form of agriculture, bird conservationists are fighting a battle for all of nature – it's just that the rest of nature lost more of the war, longer ago.

From the farmer's point of view

I've met very few farmers who actually want to get rid of wildlife from their land – but, just for the record, I have known a small number. I have met quite a few farmers who are rather indifferent to wildlife and these also tend to be those who are pretty ignorant about it too. In my experience, most farmers are quite, or very, interested in wildlife – it's not surprising that they are interested in what lives on their land and shares their place of work. But only rarely does that interest translate into the actual wish to help wildlife thrive.

The economic pressures on farmers come and go, and not all farmers are in the same economic position at the same time. But farmers have been told different things by different people for a long time. Agricultural colleges, government policy, ramblers, local communities, the vested interests of agronomists and pesticide companies, nature conservationists, supermarkets, the organic movement, the NFU, the EU and other farmers have all had their say.

Not surprisingly, farmers are at their most comfortable when producing lots of food – whether it be wheat, milk, eggs or meat. That's what they get paid for and that's what they want more of – like any other business. Much of the training that farmers get is directed towards squeezing that little bit more out of the land in the short term – an extra tonne of wheat or a few more litres of milk per cow. And that is entirely understandable – every business strives to be more efficient and more productive and to do a better job. And there can't be a much more important job than food production – we all need to eat, and very few of us produce any of the food that passes our lips every day so we need other people to do it for us.

Farmers get very mixed messages about what farming is for – and the messages have changed over time from Dig for Victory onwards. Here is my take on it. Farming is a business unlike any other except, perhaps, forestry. It is different because our need to eat is so much more essential than our need for, say, a new colour TV. Farming is also different because of the impact of the way in which it is done on the land around us and everything that that entails. Farming practice affects the landscape, the water that we drink, the greenhouse gas emissions that will alter our climate, the wildlife around us and the lives of every town and country dweller alike.

Farmers can't easily sell us the song of a skylark or soil carbon for real money – even though we value those things (or should do as individuals and as Society as a whole). So it is not surprising that over time the decisions of individual farmers, influenced by all those voices telling farmers what they should do, have tended to favour the money-making options over the wildlife-protecting options. It's not surprising that wildlife has lost out. It's not really farmers' fault – although they are not completely blameless either. We need to find a better way forward.

The cost of food

It's a fact that the average household spends less, as a proportion of its income, on food now than in the old days. We are richer and the price of food has become more affordable for us all in the wealthy west. And, a good proportion of our food is nowadays produced abroad and not by our farmers – and that applies to lamb (from New Zealand), apples (from France), as well as to Kiwi fruit and oranges (from Israel). But our farmers are exporting their produce around the world too. It's a global market and our higher land, labour and fuel prices mean that our farmers are increasingly forced to be more and more efficient in order to make a living.

These days, an arable farmer in East Anglia must be as adept at making decisions on when to sell his or her wheat on the Chicago futures market as to when to plant, spray or harvest it. But at least that farmer is, as I write, selling his wheat at a much higher price than a few years ago. In the shorter term, on the day I write these words, wheat prices are around £144/tonne whereas a few weeks ago they were at £170/tonne. If you sold then, you are smiling – if you held on then, it was because you are hoping for a future upturn in prices.

But these days, a dairy farmer in the West Country will be struggling to sell his milk for much more than the cost of production and life is hard. A farmer's life is hard not just because it is tied to the land (particularly if you have livestock), at the mercy of the weather, diseases and pests, but also because of the economic difficulties of getting a fair price for produce from the supermarkets, or the markets generally.

For the consumer, however, food is cheaper than ever – or is it? We actually pay for food several times and in several ways. We've already made a contribution before we go into the shop and fill our baskets and trollies thanks to that huge edifice, the Common Agricultural Policy. The CAP is a way of transferring money from the taxpayer to the farmer. There are lots of taxpayers and many fewer farmers (who are of course taxpayers themselves) so a little bit from each of us makes quite a large cheque when it arrives through the farm letterbox.

The history of the CAP is long, complicated and not very glorious and it is a story of good intentions only partly realised as good outcomes. It started when the European Economic Community was a small club of six nations (France, Germany, Italy, Luxemburg, the Netherlands and Belgium) and now exists in an evolved form in an EU numbering 27 members spread from the Azores to the Arctic.

The CAP has always been mostly about food and partly about farmers and rural communities – but increasingly it has become a little bit about the environment too. It was about subsidising food production so that a producer would get a guaranteed price for their wine, olive oil or beef regardless of the actual market demand for it. The remnants of those payments persist but are not now linked to production – they are payments just for being a farmer (and are based on your previous payments under the old system so the 'shadow' of production subsidies persists). These payments come with very few strings attached – non-farmers often find it difficult to understand the largesse that they represent. They are payments for being a farmer and they form an important part of farm economics – they are simply income support. But unlike means-tested state benefits these payments go to rich and poor, needy and comfortable, good and bad farmers alike.

There are few payments of public money which are so untargeted in their distribution (child benefit might be another) and we aren't talking small amounts of money either. The 'just because you are a farmer' payments to farmers in England and Wales are in the order of £1.5bn per annum. Given that these are payments without conditions they don't achieve much except transfer money from you to a farmer. In fact they provide a cushion for farming to be both inefficient and uncaring because they don't encourage good practice or environmental stewardship. The old subsidies unwittingly encouraged environmental harm but their new incarnation does nothing to encourage environmental good.

There is a part of the CAP payment which does support farmers who want to do environmental good – it comprises £0.5bn of the total support of £2.0bn – and we will come back to it later as it provides some hope for farmland birds and other wildlife in this complicated mix of world trading markets, EU hand-outs and input prices.

So, you pay for your food in the shop and you pay for it through your contribution to the CAP, but you do a bit more paying as well. You pay through what economists call externalities. The externalities of farming are those things that I described earlier which make farming different from widget-production – landscape, wildlife, water, greenhouse gases. Some are good and some are bad and in theory, and sometimes in practice, we could put a cost on each one. We could cost the beauty of the countryside in terms of whether people would pay to preserve it or whether they act as though they like it by visiting it and spending their cash in the local petrol station or shop. The local farmers might say that they are creating the product that creates wealth for local businesses – and as wildlife is part of that natural beauty then we could lump wildlife in there in the calculations.

Some of the costs are a good deal easier to calculate – run-off from agricultural land creates water quality issues that need to be corrected by water companies through treatment works. Treatment works cost money to build and run and these costs appear in your water bills, so the more polluting agricultural practice is, the higher your water bills and those of the farmer as well. And this is a price you pay for how your food is produced and money that the farmer does not see. The increased cost of water bills due to treating agricultural pollution is about £300m per annum for England and Wales – and that is a tidy sum. In addition, there are the externalities arising from soil carbon and nitrogen fertiliser lost into the atmosphere, both potent greenhouse gases, as well as many other factors that could be quantified and valued. The point is that agriculture is not just a business like any other. It has wider impacts on our lives and the decline of skylarks in the fields where I walk are but one symptom of a huge edifice of economic, social and environmental damage, quite unlike those of any other industry.

That's what makes farming different and it is the justification for public money going into it as an industry rather than just saying to farmers 'make your money from the market'. But supporting the continuation of public funding for agriculture is not to say that the current investment is well-targeted or that it gives good value for money – and it is not to say that we should not press for improvement.

A European problem

I write this book, and this chapter, at a time when the future of the EU is more uncertain than ever before. The CAP is an EU-wide policy that influences land use in 27 countries across the European continent. The UK has a say in the future of the CAP, but only to an extent. As the EU has expanded from nine to 27 countries, the task of making good decisions has become much more difficult. Imagine that you meet up with 26 close friends and are deciding where to go for dinner – how easy would that relatively trivial decision be? Well, making changes to EU policies has become rather like that – and these 26 countries are not all good friends.

High level agricultural policies are made at an EU level. In a devolved UK, the four countries will each have a say in what the UK position will be but then one UK Agriculture Minister will meet with his or her counterparts from each Member State to make changes to the CAP. It is the job of nature conservationists to try to influence the position of the UK in those talks, not just to protect UK nature but EU nature as a whole. And so RSPB staff meet regularly with our European counterparts to discuss and agree our positions on the issues of the day – including our close friends in the BirdLife International family (the lead organisations for each country being the LPO for France, the SEO for Spain, and so on) as well as international organisations such as the World Wide Fund for Nature (WWF) and others.

Although the discussions between nature conservation organisations in Brussels are probably a lot friendlier than those between Heads of Government, they sometimes have their moments and even some of the tensions that exist to a much greater extent at governmental level. There are obvious differences in language, wealth, knowledge, ability and local circumstances. But overall, the BirdLife partnership works very well as a group and its small secretariat based permanently in Brussels acts as our eyes and ears, and often as our joint mouthpiece too. The work of BirdLife International on influencing the policy agenda in the EU is a model of international cooperation and all I would say is that if we had been in charge of all the EU's decisions they might have been made a lot more quickly than under the current arrangements.

Working with farmers

I love farmers – some of them. And some of them love me – although some of them don't, and that's OK too. While I was at the RSPB I made a lot of farming friends who got to know me and to whom I talked a lot, and from whom I learned a lot about farming. I can't remember ever turning down an invitation to talk to a group of farmers unless I really couldn't make it and I spent many a winter evening in a village hall or pub meeting room talking about farmland birds, getting told off by the most vociferous farmer in the room, and then making the long drive home. Driving home from a not very uplifting meeting in deepest Essex, I recall thinking to myself that I had been in my office at The Lodge just before eight in the morning and had put in eight hours before leaving for Essex only to get home just before midnight. So I'd done the best part of a day's work for the RSPB and then put in another day's work in order to be abused by the farming community as their guest (and the RSPB paying for the petrol too!). That evening I wondered whether it was worth it but on other evenings it was absolutely great. I recall evenings with farmers in north Essex/south Cambridgeshire where after robust and friendly discussion and argument, a meeting of minds followed and I went home with plenty to think about and feeling very pleased with having spent the time.

There are lots of great farmers out there – people like Rob Law, Duncan Farrington, Patrick Barker and Henry Edmunds – just four of many farmers I met and got to know through the RSPB.

Rob is a main supplier for Jordan's Cereals and farms down near Royston. He's a big man and always willing to talk and debate in a very friendly way. He has roped me into giving lots of talks to farmers' groups and I've never regretted them because he sets the right tone of friendly challenge. He looks like a rugby player and if he were, he'd be the type whom you would be very happy to have a drink with in the bar after the match.

Duncan farms near where I live and I first met him when I volunteered to do a bird survey on his farm. He soon rumbled who I was and I've kept in touch with him and worked with him on Open Farm Sunday since. When I surveyed his farm, which I had driven past hundreds of times, I found turtle dove (a Northants rarity these days) and a couple of pairs of lapwings too, as well as plenty of yellowhammers and linnets. Duncan's is not only a Linking Environment and Farming demonstration farm but he also sells cold-pressed rapeseed oil called Mellow Yellow. The Mellow Yellow mayonnaise is particularly delicious – so buy some and support this wildlife-friendly farmer.

Patrick Barker farms in Suffolk and knows his birds – including his turtle doves which are being studied by the RSPB. I'm as likely to see Patrick at a birding event as at a farming one. One of the breed of young farmer who I hope might replace some older dinosaurs.

Henry now farms organically down in Hampshire and has stone curlews and chalk grassland on his farm because he looks after them both. Visit the Cholderton Farm Shop and help him continue to do so. He's a great character and delights in showing what he has achieved for wildlife on his farm – it's a lot and I know it has been hard-won over the years because I know some of the back-story.

I could go on, but these are just four different farmers whose paths have crossed with mine over the years and with whom I have spent time because they are reasonable and sensible and entertaining people who are doing a difficult job well, in my opinion.

There are lots more like them out there but not quite enough because farmland wildlife continues to decline. That's not the fault of those who are doing the right thing but those who are aren't numerous enough to compensate for those who aren't. And you only have to drive around the countryside to see plenty of farms with awful hedges and crops grown right up to the field margins with not even a nod towards wildlife. Wildlife-unfriendly farming isn't a secret – you can see it everywhere.

Helping the good guys is always a good idea and during my time as Conservation Director the RSPB increased the resources it put into working with farmers. We spent more of our money, your money if you are an RSPB member (and I hope you are), on farmland advisors to work with wildlife-friendly farmers across the UK to get a good deal for birds.

We also started a very successful award for good farmers called the Nature of Farming Awards which attracts a large number of entries and the winner of which is decided by a public vote from a shortlist selected by experts from the RSPB, Plantlife and Butterfly Conservation. This seemed a great way to recognise the best farmers and it was encouraging that it soon became the most sought after farming award – judging by the number of entries anyway.

But one of the farmer-friendly projects of which I am most proud is the Volunteer and Farmer Alliance where willing volunteers are matched with willing farmers and the volunteer surveys the farm and tells the farmer what birds he (or she) has got. This was how I met Duncan Farrington – by being his volunteer. But I also remember getting the then President of the NFU, now Sir Ben Gill, to enter this scheme. I was giving a talk to a group of horticulturalists in London, at Ben's request, and showed a photograph of a stunning bright-yellow male yellowhammer. On the spur of the moment, and with no malice in mind, I turned to Ben and said 'You know what this is, don't you Ben?' But unfortunately he was stumped. I hadn't meant to embarrass him, and to be fair, I looked more embarrassed than he did, but I followed up with 'What you need is a survey from our V&FA project' and so he said yes. That spring a volunteer visited Ben's Yorkshire farm and found not only lots of yellowhammers but tree sparrows too. And every time I see Sir Ben these days he tells me, with pride, about how the tree sparrows and yellowhammers are doing.

I tell that story not to re-embarrass Sir Ben Gill, but to make the point that farmers often don't know that much about the wildlife on their farms. Certainly living in the countryside all your life and farming there are not guarantees of wildlife knowledge. Many farmers don't get past the big species of hawk, pheasant and crow in their bird identification and it is unusual, though all the more pleasurable when it does happen, to go out with a farmer who knows the simplest of bird songs.

Sir Ben was the President of the NFU and I've met and worked with all of his successors. The NFU is a powerful force in the world of agriculture and it is a union for farmers. It isn't there to be nice to wildlife – which is just as well because I can't remember it being nice to wildlife for a long time.

Being a farmer – the Hope Farm experience

In the late 1990s the RSPB decided that farmland birds were such a big deal for us that we needed to do something more for them. We came up with the idea of buying a bog-standard arable farm near to The Lodge, so that it could be easily reached by our staff and was only a little over an hour from London by train. We put our money where our mouths had been and set out to demonstrate that profitable farming and growing bird numbers could go hand in hand.

And so we started looking for a property. Roger Buisson was given the job of finding the right farm and the right farm had to be a normal one – not a 'lost in the mists of time' farm or somebody's hobby farm but a real farm that any farmer would recognise as a real farm. And it had to be within budget – a sum I shall not disclose! After one significant false start we found a farm about 20 minutes drive from Sandy and near to Cambridge. Hope Farm, as it came to be known, was on the small side (but that made it cheaper) and was certainly ordinary. As part of the process I took a series of RSPB people to see it and Ian Newton, Chair of Council at the time, was among them. He was keen on the idea and was already planning which ways to manage which bits of the farm. On another occasion I took the RSPB President, Julian Pettifer, around the farm and he, too, was very keen. I knew that my boss, Graham Wynne, would be one of the more difficult to persuade – that was his job after all!

But we took the proposal to the RSPB's Council and they were keen on the idea so then it was all systems go. First, we had to try to raise some money for this expensive purchase and we really did wonder what RSPB members would make of the project. Appeals to buy nature reserves were almost always well-supported but we knew that this was a very different prospect. For a start it wasn't a site that we could encourage people to visit. It was a working farm in a small village with no easy access and, being a normal farm, it had all the wildlife of – well, a normal farm. But more importantly we couldn't claim that buying this farm would save a species or even that we would definitely be able to use it as a lever to move the world. It was a different and ambitious project and it is greatly to the credit of RSPB members that they took it to their hearts and supported it to the tune of over £1m.

Hope Farm is 184 ha and apart from a few small paddocks is wholly arable with large fields and unimpressive hedges as field margins. It had been owned by the same family for decades and they were able to tell us something of its history. The story of Hope Farm was a microcosm of the changes that had occurred in British farming in this part of the world. Hope Farm was indeed pretty normal. Back in the 1950s it had kept livestock and had been called Dairy Farm. As time went by the dairying was lost and although some beef cattle were retained the emphasis changed to arable farming until by the 1970s the farm was solely arable. Hedges were removed to create big fields to allow block cropping and the number of people employed on the farm fell as did the variety of crops being grown. When we took over Hope Farm the crop rotation was two years of wheat followed by a year of oil seed rape and the owner was seriously considering moving to growing wheat every year in every field. Set-aside was still EU policy and so a small proportion of the cropped area had to be rested every year, but on Hope Farm this area was used, as it could be under the rules, for growing more oil seed rape for biofuel use.

The first year at Hope Farm was a busy and fascinating one. It was a 'getting to know you' year with baseline surveys of birds, bugs and plants and lots of planning for the future. When you extend a nature reserve you often know what wildlife you are getting – you've looked over the fence. When you buy a new piece of land for habitat recreation then you set out to create the new habitat and watch the wildlife move in. But this was very different as it was a normal farm with very little legacy information on its wildlife. We had bought the farm in winter and as the spring unfolded, its bird potential did likewise.

The site certainly had lots of potential as it didn't start with many birds. There were skylarks, yellowhammers and linnets in fairly small numbers but no grey partridges, lapwings or yellow wagtails.

Initially we contracted Cooperative Farms to do the farming but after a few years they moved their operations and we found a local farmer to take over the contracting. In practice, the contractor is paid a fee for the work required and then, as an incentive to farm well, he takes a share of the profits too. So the RSPB could rely on the farming nous of our contractor and concentrate on specifying crop and management practices that might encourage wildlife without wrecking the farm income. One of the first decisions was to put the set-aside allowance back into proper set-aside rather than industrial oil seed rape. This would restore the equivalent of stubble fields in the winter as a much-needed source of food for farmland birds and provide more nesting areas for ground-nesting birds such as skylarks. In time we switched the crop rotation to a four-year cycle of wheat, beans, wheat and rape. We introduced skylark patches into our wheat fields and saw a few of them fill up with weeds, but the fields fill up with skylarks again. We experimented with a two-year set-aside, which worked well for birds but proved tricky for the farming – and then set-aside payments were done-away with anyway. We introduced some small water features to investigate their impact on removing nutrients from the water running off our land. And we paid a lot more attention to the price of wheat and rape seed than we had before and to the rising costs of fertiliser and pesticide inputs.

I cannot stress enough that Hope Farm was, and is, run as a farming business. Clearly the whole of the RSPB wouldn't fall if Hope Farm weren't successful but this wasn't play-farming, as most of the farmers who visit Hope Farm each year acknowledge. Herbicides were used on all the fields in just the same way that neighbouring farmers used them and although we were probably more cautious about using insecticides, and followed best practice by leaving unsprayed headlands, we still sprayed in those years when orange blossom midge was a real problem for the wheat crop. This is farming that makes the same economic decisions as are made by our neighbouring farmers. And the RSPB's books are independently audited and the success or failure of the farming at Hope Farm is there for all to see.

Although this farm was not open to the public like most RSPB nature reserves, we did invite people to see it. Most years there were visits from local RSPB members' groups and occasional organised visits from RSPB members, like myself actually, who had contributed to the costs of buying the farm. These were fascinating events as are all those involving RSPB members because members are, all one million plus of them, different. So a visit to this normal farm in Cambridgeshire would include a few keen birders, perhaps a retired Ministry of Agriculture, Fisheries and Food civil servant, a farmer or two and a lot of other very bright and perceptive people – they were enthusiastic groups. I remember one weekend when just about every visitor got soaked the weather was so awful – which meant the staff involved got soaked four times (each morning and afternoon on Saturday and Sunday). But the RSPB members were still thrilled to see what their money had allowed us to do and a brief view of a singing skylark made the day for many. Most of the visitors went home happy, and some had travelled a long way for the visit, and staff went home enthused by the enthusiasm of the people who paid our salaries.

But there were also many other visitors, including hundreds of farmers, who would look more closely at the crops and then at the birds and then put detailed questions to a series of excellent farm managers. Roger Buisson, Darren Moorcroft, Chris Bailey and now Ian Dillon all had to put their ecological understanding to the test with some audiences and their agricultural understanding to the test with others.

Hope Farm has been visited by MPs and Ministers, by Westminster civil servants and Brussels civil servants, by NFU and Country Land and Business Association Presidents, and by a host of land agents, agronomists and land managers.

In the early days we told Hope Farm visitors a story of ambition and our sticking our necks out – mostly my neck it felt like sometimes. Our goal was to maintain the productive nature of the farm in comparison with neighbouring farms but at the same time to boost its wildlife content. Could we do the latter whilst doing the former? We thought we could, and hoped we could, but now we had to do it.

Farming like a normal farmer meant that the options for change weren't enormous. In the first years we did reinstate set-aside and you could see that fill up with birds straight away. We trimmed the hedges less often to allow more berries to form and we were probably, although it's difficult to tell, just a little less harsh with lots of small farming decisions. We changed the crop rotation largely on agronomic grounds to combat arable weed problems, but the introduction of beans into the rotation was almost certainly good for birds. Hope Farm had more skylark patches than any other farm in the country for many years as we tested their impact and showed that they did increase skylark numbers.

Our main ally in putting nature back into Hope Farm was that part of the enormous CAP budget that is available to farmers across the EU (generally speaking) and to farmers in England, to assist wildlife-friendly farming. In England, now (things have changed over time), there are two bits of the scheme – the Entry Level Scheme (ELS) and the Higher Level Scheme (HLS). The HLS is quite difficult to get into and is aimed at those landowners who want to go out of their way to help nature on their land. It is really good value for money and very effective but it isn't for everyone and at Hope Farm we were trying to show that nature and farming could fit together without either having to bend over backwards to accommodate the other. So the ELS was the option we followed. Any farmer in England can join the ELS option provided they clock up enough points for wildlife-friendly farming. Once you pass the threshold you get the payment, but if you do more good things for wildlife you don't get more money. It's a threshold scheme and you either are in or out, and if you are in you get a certain payment from the UK government and the EU – which means from you the taxpayer, really.

This scheme was designed as a broad and shallow scheme following the foot and mouth years. It's supposed to be easy to do but is supposed to be effective too – after all your money is paying for it. Not unreasonably, most farmers look at the options available and pick the ones that suit them the best and are the least trouble to implement. When I say 'least trouble' that may give the impression that they are some trouble but many farms easily pass the points threshold for ELS entry without having to break sweat. And that's why the ELS isn't working well enough at the moment – it's just a little too easy and a little too ineffective as a result. Despite around 70% of the farmed area of England being in ELS, farmland bird populations continue to decline.

Government figures released in November 2011, bringing the Farmland Bird Index up to date for 2010, showed farmland birds at their lowest levels since the baseline year of 1970. Most of farmed England is covered by a wildlife-friendly scheme and farmland birds still tumble in numbers!

We had to do better at Hope Farm and so we chose options that would work best. We put in skylark patches and beetle banks and we planted a few nectar-rich field margins as well as grass margins in some fields. It wasn't a big deal. We still produced very good crops of wheat because the field edges are almost always the least productive parts of a farm – and the area given over to wildlife moved around the farm a bit and was always a very small proportion of the whole. But even though you would hardly notice the difference if you drove past the farm and glanced over the hedge (which was a bit thicker and less trimmed) if you walked the fields and counted the birds then over time you did see a big difference.

Skylarks responded well to the skylark patches and their numbers leapt from around 10 pairs in the baseline year of 2000 to 42 pairs in 2011. Yellowhammers increased from 14 to 33 pairs and linnets from six to 26 pairs over the same period. And in case you think I am picking out the good news then the FBI for Hope Farm increased by 211% (i.e. it more than trebled) in the period 2000–2011 whilst the national FBI fell by 11% in the period 2000–2010 (2011 figures are not yet available).

Grey partridges were absent at Hope Farm when it was bought and for several years afterwards but there are now five pairs nesting successfully. This, I have to say, was always a cause of mild embarrassment to the GWCT whose own farm, at Loddington in Leicestershire, has been as successful as Hope Farm in most respects but having started with nesting grey partridges subsequently lost them. The grey partridge is, sadly, the GWCT's logo bird.

Yellow wagtails were absent from Hope Farm when it was bought but have returned and in 2011 there were two pairs. Both turtle doves and lapwings have returned to the farm, so it now has almost the complete suite of farmland birds. Even the corn bunting, which got our hopes up in 2000 by nesting but then went missing for years, returned as two pairs in 2011. The one species still lacking is the tree sparrow – and considering it was studied at the nearby Agricultural Development Advisory Service (ADAS) farm at Boxworth there must have been lots of tree sparrows around in the 1980s. The RSPB awaits their return with impatience.

The RSPB stuck its neck out at Hope Farm – it could have been an awfully public and expensive failure, but it has been a great success. Let's be clear, financially we would have been better off sticking the money in a deposit account, but the farm has been profitable and yields have been maintained at levels better than the regional average throughout. And land prices haven't fallen so the RSPB owns a valuable asset. But in nature conservation terms the asset has appreciated considerably in value with a trebling of the natural capital as measured by the FBI. Profitable farming and increases in farmland birds can make good bedfellows – and the RSPB has shown how. Hope Farm was well named.

The way forward

The scale of loss of farmland birds in the UK is bigger than elsewhere in Europe and came as a big surprise to me in my early RSPB career. The figures show the absolute value of good monitoring schemes. I suspect that other farmland taxa have declined just as drastically but the data aren't there to demonstrate it. During my career in nature conservation these declines of common and widespread birds have persuaded me that nature conservation isn't just about the rare species and those near to extinction – it is also about the decline of the formerly commonplace and familiar.

And once you get interested in the declines of common, widespread birds you are inextricably drawn into an interest in global issues and public policy in areas that seem greatly removed from birds and their biology. To save the skylark you need people in suits in meetings just as much as you need people in wellies in fields.

Over the years it has been a privilege to work with so many bright people from BirdLife partner organisations across Europe. I now have friends in Spain, France, the Netherlands, Greece, Germany, Austria, Portugal and elsewhere which I would never have had without those visits to Brussels and our joint work on agriculture and other issues.

The losses of birds in our countryside are the last knockings of massive biodiversity loss across a wide range of taxa – plants, mammals and invertebrates. Despite what some farmers will say, the declines in bird numbers are real and arguably the most striking sign of ecological change that we have seen in the UK in recent decades. And despite what some will say, it is well-established that changes in what farmers do on their land are the main causes of these changes in wildlife richness. That isn't to say that we should blame farmers for all of this, although we shouldn't let the farming community completely off the hook either.

The way forward is quite clear if we want our farmland wildlife to recover. We know that well-designed agri-environment schemes will deliver more wildlife in our countryside and that this can be done with very little impact on farming profitability. However, leaving a bit of a space for wildlife will impose some cost on farmers and since wildlife is a public benefit, that benefit should not become an imposed cost on the farming industry. However, we taxpayers are already pouring money into farming in a very inefficient way. It's inefficient in terms of income support, as most of the money goes to farmers who don't need it, and it's inefficient in terms of environmental delivery, in that it pays farmers to do things that don't work well enough, when we know that better ways exist. And most environmental delivery is through voluntary schemes which are not very efficient either.

If we want some wildlife back in the countryside then the current system of payment to farmers needs to be overhauled – the same amount of money could buy us far more wildlife. But if we, as a society, aren't bothered about wildlife then let's just have our money back and invest it in computer games, nurses or alcohol.

A two track approach is needed – working with farmers who are warm to wildlife and working with decision-makers to make the whole system more wildlife-friendly. There are too many wildlife-friendly farmers to ignore them but also too many to be able to work closely with all of them. And there are too few as a proportion of the farming community to rely on them for the recovery of farmland birds and other wildlife – a perfectly feasible goal armed with the knowledge we have.

> *O greatly fortunate farmers, if only they knew*
> *How lucky they are!*
> **Virgil**

CHAPTER 8

Reintroductions: putting something back

Shooters 'Say no' to Suffolk sea eagles
Shooting Times headline

You can sometimes feel a bit powerless as a nature conservationist. When you realise that the absence of skylark song from fields where you once listened to them is influenced by world trade agreements, the pressing need to feed a growing world population and complex policies set by EU countries, you might be forgiven for fearing that it's too difficult a problem to tackle. Experience proves that view to be partly right and partly wrong, but one of the big advantages of working for the RSPB was that it excels at practical conservation work. And so, when the big policy areas seem to be intractable and you want to stop banging your head against the CAP, or you've seen enough of the NFU President for a while, then you can always take pride in the practical actions that an organisation like the RSPB can take without asking the EU's permission.

One of the more uplifting practical measures, alongside buying up Britain (which we will come to in the next chapter), is reintroducing lost species. Species reintroductions are often controversial, sometimes with those who don't work in nature conservation but also sometimes with those who do. I think that is a shame. Reintroductions have always seemed to me to be the species equivalent of habitat recreation. Rarely do people complain on principle about putting back a reedbed or a heathland in areas where they used to be (they might argue about the details, the location, the cost, the methods and almost everything else, but not usually about the principle) and I've never seen much difference between that and putting one particular species back into a place where it used to live.

One of the arguments against reintroductions is that they are expensive but I don't think that they are – provided that they are successful – and this argument is usually advanced by those who have set their mind against a particular project on other grounds. But since we are talking practical action, let's consider some actual projects, rather than the theory.

Red kite

When I grew up in Bristol you had to go to mid-Wales to see red kites. Mid-Wales had other things to offer too – oakwoods with pied flycatchers, wood warblers and redstarts, streams with dippers and common sandpipers, and hillsides with ring ouzels, ravens and whinchats. But kites were a great prize. I remember the family holiday to mid-Wales when I saw my first red kite, in the far distance, on the last day of the week-long holiday. I was delighted to be able to pick out the red forked tail on that far-away bird hanging over an oakwood on a steep Welsh slope. And a few years later, on a weekend away with the Bristol Grammar School Field Club, we walked over the newly coniferised sheepwalk around a now-defunct Youth Hostel with black grouse bubbling away at a spring lek and a beautiful red kite circling overhead. Wales was the place for kites. It seemed like their natural home.

But others realised that this was actually a sorry state of affairs and years later when I started working for the RSPB plans were afoot to reintroduce red kites to both Scotland and southern England. The discussions were well above my pay-grade at the time, and that's probably just as well, because I vaguely remember being pretty nervous about whether releasing red kites into the Chilterns would work. Surely a species that thrived up a hill in deepest Wales, apparently feeding largely on dead sheep, wouldn't do that well amongst all those posh southern folk where Thelwell ponies were commoner than sheep and large pheasant shoots were everywhere. Would the Chiltern gamekeepers let kites survive?

The answer is now very clear – red kites did brilliantly in the Chilterns and those (in the Nature Conservancy Council and the RSPB) who made the decision to go ahead were thoroughly vindicated by the huge success of this reintroduction scheme. Illegal persecution, including poisoning, proved to be much more of a problem in the north of Scotland release site, on the Black Isle north of Inverness, where the growth and range expansion of the kite population has been much slower. Building on both schemes, a successful network of red kite reintroduction sites was set up. Up and down the length of the UK and into Northern Ireland red kites are now so common that their numbers can be monitored by the Breeding Bird Survey and they don't need special kite-focussed surveys any more.

What the architects of the red kite reintroduction schemes knew was that mid-Wales was the last place that kites had hung on in the UK after persecution had swept them from most of the rest of the country – but it was also one of the last places that a red kite would choose to live were it not molested. The much richer lowlands with their small mammals and earthworms, and warmer climes, were much more attractive to red kites than a living made on rotting sheep carcasses on a wind-swept hill in Wales (no offence intended to the land of my mother).

When the next Atlas is written and published take a look at the map for red kite and see how much it has changed since the last Atlas – this is a landscape-scale species conservation success story, for sure. And as I write these words I can look out of my Northamptonshire window and stand a good chance of seeing a red kite over the garden – and each time I do my heart lifts.

I used to use the red kite example in talks to RSPB members' groups and other birdy audiences. One of the stories I would tell was about the fourth Baron Lilford, of Lilford Hall, Northants. Lilford Hall lies just up the road from me and I drive almost past it to catch or meet a train at Peterborough. The fourth Baron Lilford was a remarkable figure in birding circles in the late 19th century – a founder member of the British Ornithologists' Union, the person who successfully introduced little owls into Britain (of which more later) and the author of a two-volume work on the birds of Northamptonshire. He also had a large collection of captive birds, including flamingos, and the story goes that there were at one time free-flying lammergeiers over the Jacobean-style Lilford Hall and its parkland.

In his book on the birds of Northamptonshire, Lilford regretted the absence of red kites from the county at the time and reminisced about being taken out onto the lawns of Lilford Hall, as a boy, to be shown three circling overhead. Even in the late 19th century red kites had vanished from the Northamptonshire skies and yet now they are back in numbers. As I drive past the Lilford ancestral home these days I am more likely to see a red kite than not. And I think it is a sign of great progress, a fantastic conservation achievement, that my children and their school friends belong to the first generation of kids in 150 years to grow up in Northamptonshire and have red kites back in their lives as a commonplace. If their baselines for farmland birds are much lower than mine at least their baselines for red kites are higher – and it would not have happened without active reintroduction.

I can't resist telling you another Lilford red kite story even though it has nothing to do with reintroductions. Lilford was a keen visitor to Spain – to shoot and to watch birds and other wildlife. He tells the story in his *Birds of Northamptonshire* (where it is as irrelevant to the theme of the book as it is here) of being in central Spain and climbing a tree near Aranjuez, south of Madrid, to visit a red kite's nest. He makes the point that kites collect lots of weird things in their nests, and they still do – a colleague was always talking about finding women's knickers in a kite's nest (presumably plucked from a washing line rather than from a woman). So you have to picture this aristocratic Englishman, dressed I always imagine in tweeds, clambering up a tree in the plains of central Spain to look into a kite's nest, in late April 1865. And we can be sure of that date because Lilford tells us that it was on a scrap of newspaper he found in that kite's nest, up that tree, that he first learned of the assassination of President Abraham Lincoln (which occurred on 15 April 1865).

I told that story at a red kite event at the FC site of Fineshade Woods in Northamptonshire, where there are now red kite walks on which you can see up to 100 kites at a winter roost. But another story worth telling is that at Fineshade sometimes a CCTV link is set up looking into a kite's nest, giving amazing images of this bird at the nest. A group of slightly sceptical local gamekeepers were once taken to see the live video coverage and learn about kites, and as they watched a parent arrived at the nest with a dead stoat in its claws – we couldn't have planned it nor believed it possible. The group of 'keepers looked amazed and then broke out into a spontaneous cheer!

At work at The Lodge, my first red kite record was on 10 April 1997 when Gwyn Williams and I saw one as we headed off somewhere together early one morning. Back then a red kite sighting might cause dozens of people to leave their offices to look whereas now, and this is a good thing, it is a fairly regular occurrence and causes less excitement. And I am reminded that some birders argued against this reintroduction project because it would mean that you wouldn't know whether a red kite you saw was 'real' or not – inexplicable to me.

One of my favourite local eating and drinking places is the Snooty Fox at Lowick where in summer, perhaps after seeing black hairstreak or purple emperor butterflies nearby, one can sit outside and see whether the kites or buzzards are in greater numbers overhead as one sips a glass of beer. And if you cannot get to Lowick then why not try the red kite ale produced by the

Black Isle Brewery – situated in one of the two early release sites?

I remember doing a favour for a local farmer on Open Farm Sunday and telling visitors to his farm about the birds they might see. A young boy was bubbling over wanting to tell me about the kites that he sees from his school playground and how great they are. His parents looked on proudly and Dad said to me as they moved on that he really liked the fact that the kites meant so much to his son and that they were now back above the playground, while his own childhood had been kite-less.

One of the reasons for telling these tales of red kites is to make the rather obvious point that if they weren't now common in rural Northamptonshire and beyond then they wouldn't form a part of my life and my life would be slightly the sadder for it. If wildlife is absent then we do not miss it because we do not know it. When it returns, it enriches our lives and we realise what we had been missing before. And reintroductions can be a very good way of putting that richness back into the landscape. If the RSPB were looking for a new logo to sum up conservation success and a message of hope, which it isn't, then the red kite would be a strong contender.

Corncrakes

I saw my first ever corncrake on another family holiday – this time on Islay. It was a trip that produced my first ever chough on the Oa, which is now an RSPB nature reserve, and great views of hen harriers by the side of the road. But corncrakes were to be heard craking from the fields in those August days of 1972 and we once stopped our car to allow a brood of young black chicks to scurry across the road with their parent. It seemed, just like Wales was the place to see kites, the Hebrides and Outer Isles were the places to see corncrakes. But – just as red kites had once been found throughout the UK – so had corncrakes. Corncrakes were a familiar farmland bird at the end of the 19th century and then nested in every county in England, Northern Ireland, Scotland and Wales. Their loss from virtually everywhere except the Western Isles and Orkney during the 20th century was mirrored in many other European countries with 'efficient' agricultural systems and for quite some time the corncrake was listed as being in danger of global extinction.

Several people associated with the RSPB played crucial roles in understanding the decline of the corncrake. Perhaps the first of these was Tony Norris who died in 2005 and whom I never met. His studies of the corncrake (including an enquiry which he did for the BTO at the age of 20 in 1937) documented its loss and the causes of the decline. Much later, folk like James Cadbury, Tim Stowe, Glen Tyler and most notably, Rhys Green, did the work which revealed the causes of the corncrake's decline and, most importantly, pointed the route to its recovery. Put incredibly briefly, corncrakes declined because they are short-lived and need to be very productive each year to keep the population going. They nest through into late summer in crops, notably hay meadows, which are now harvested earlier than they once were, and their last broods tend to fail through their being scrunched by harvesting machinery. The reason that corncrakes hung on in the northwest of Scotland was that harvesting was later there because of climatic and social conditions – slower growing crops and less mechanisation.

Once the reasons for decline were fully established solutions could be deployed. Nature reserves were established to provide havens where agriculture could be tailored to corncrakes rather than hoping that corncrakes would adapt to agriculture. Incentives for farmers – usually, like the St Kilda and Shetland wrens, that farming sub-species called crofters – were established to delay hay-cutting in corncrake areas and to cut fields in ways that allowed corncrakes a more sporting chance of escape from the harvesters. These conservation measures were deployed as the full-package RSPB response – advisory work with land managers, buying nature reserves across the corncrake's remaining range and influencing grants to farmers to encourage corncrake-friendly farming practices. And it has been a great success – someone else could write a brilliant book entirely about that story and how the corncrake's long term and precipitous decline was halted – and how some said that it couldn't be done. But that is another story. The corncrake's successful increase in the Western Isles and Inner Hebrides has enabled us to stabilise a much-reduced population. It would take much longer to get corncrakes back onto southern English farmland, where they were once very common, so some of us turned our minds to reintroductions.

Corncrakes need grasslands to nest in – and they don't thrive in grasslands which are cut early for silage or hay. Although the west of Britain is where most of the grass is these days it is also where most of the grass is cut too early for corncrakes to breed successfully. And so, perhaps on the face of it rather perversely, those given the task of thinking about corncrake reintroduction sites put the RSPB's Nene Washes nature reserve in Cambridgeshire at the top of their list. This site has many obvious things going for it – it is a large area of grassland whose management is controlled by a conservation body interested in corncrakes. But, somewhat counter-intuitively, the fact that it is a grassland island set amidst an arable Fenland desert of wheat, sugar beet and vegetable crops was also a big advantage because it would help to prevent corncrakes leaking out into the surrounding countryside. Corncrakes fly off to Africa and back before they breed in their first year and although they return to close to their natal site they do wander and spread out a bit. The advantage of the Nene Washes was, we believed, that any corncrake returning to more or less the right area would only find suitable grassland at the Nene reserve. A hostile arable area around the reserve would effectively funnel birds towards it, whereas if there were areas of less suitable grassland nearby, the population might leak out into places where they would face all the threats that had wiped them out in this area decades before. A large grassland island was perfect and since the Ouse Washes too often flooded in summer, the less flood-prone Nene Washes were the best bet.

But where do you get your corncrakes from and how do you release them? A clue to how a corncrake reintroduction might be performed came from the story of a German scientist who had been studying them in captivity. They bred so well in captivity that he had an embarrassment of corncrakes at the end of the season and simply had to let them go thinking he would never see them again – and maybe he never would because corncrakes are very difficult to see even when they are common in an area. But the next spring his locality rang to the 'craking' of corncrakes – the released birds had flown to the other side of the Sahara and back to sing near their release site. If we could do the same with many more corncrakes then we were on the road to success.

A consortium was set up involving Natural England, Whipsnade Zoo and Pensthorpe Conservation Trust to establish captive-breeding populations, to transport the young corncrakes at suitable times to holding pens on the Nene Washes and then to release them. Sounds easy doesn't it? In the process, we have learned even more about corncrakes and over the last few years scores of young corncrakes have been released each year and up to 21 calling males have been counted 'craking' on the Nene Washes. Although the project has been ground-breaking in many ways it is only at the not-having-failed stage, rather than having succeeded at the moment. Will the Nene Washes corn-crakes ever do well enough to maintain their population and fill that large eastern grassland with nocturnal craking noises without continued releases? We'll see. And if the project fails, then I'm happy for everyone to blame me for championing this bold initiative. If it succeeds then an awful lot of people deserve the credit for having got the details right over a long period of time and on a pinched budget.

But at the moment I will continue to head to the Eldernell car park on late April and May evenings to relish the harsh song of the corncrake – another sign of spring along with the cuckoo and the warblers.

In late May 2010 I visited the Nene Washes with local artist and friend Carry Akroyd and avoided the Eurovision Song Contest to boot. Almost as soon as we arrived at Eldernell and walked west we heard a corncrake (at 8.15pm). Carry and I are both big fans of the Northamptonshire poet John Clare (1793–1864), who wrote about the land rail or corncrake. It was clear that the land rail, easy to hear but elusive to see, was a familiar but mysterious bird to Clare and his rural friends. He wrote that it was 'heard in every vale' and was 'like a fancy everywhere'. And so it was for another 50 years after Clare's death when in the early days of the 20th century it was still a familiar, if sometimes irritating, noise on the evening air.

I sometimes used to think of corncrakes as I passed Tempsford on the A1, a few miles north of Sandy. There, in the valley of the Ouse not the Nene, the first book on the birds of Bedfordshire published in the early 1900s noted that the corncrake was widespread and kept the residents of Tempsford awake with its incessant calls. Let us hope, as I do, that corncrakes might more often keep people awake in southern England.

Cirl buntings

While the red kite is a stunningly attractive bird, and the corncrake is shrouded in folklore, the next reintroduction project I will consider concerns that dull little bird, the cirl bunting. I can imagine the snarls with which that last sentence was read by some of the cirl bunting's greatest devotees, but it's a bit difficult to wax lyrical over this bird. A small passerine with habits similar to a yellowhammer and a song which isn't too different either, the cirl bunting is not a species of legend – but it did decline enormously during the 20th century despite being a species that might have benefited from global warming.

I saw my first UK cirl buntings in south Devon, which now forms most of their range, as I kept missing them as a kid at their last few Somerset locations at Crook Peak and Cheddar Reservoir. This is a species which is quite common in the olive groves of Spain but also used to inhabit the fields around English villages at a time when an olive would be hard to find in your local supermarket – indeed if there were any supermarkets because the days when cirl buntings were common were also the heydays of the corner grocer shops. In the 1950s cirl buntings were widespread across most of southern England and into the Welsh borders.

Again, a very successful conservation project had identified the causes of the decline of this species – Andy Evans and Ken Smith deserving much of the credit. It needs good hedgerows to nest, good insects to feed its young (especially grasshoppers) and winter seed supplies to get through the winter. The move to more intensive agriculture had gradually removed one or other of these three essentials from counties such as Berkshire, Sussex and Hampshire until cirl buntings were restricted to south Devon where mixed farming survived in wooded valleys and on steep slopes. Again, once the reasons for decline and the package for recovery were identified, the RSPB leapt into action to ensure that the right options were available for farmers to maintain farming systems that suited themselves and suited cirl buntings.

This was another good example of the RSPB working closely with farmers. Advisory staff like Cath Jeffs spent ages chatting up farmers and filling in their agri-environment forms to maximise their chances of getting grants to retain over-winter stubbles and to do all those other things that make farmland suitable for this now very scarce bird. All this work led to a great increase in cirl bunting numbers – from 118 pairs in 1989 to over 800 pairs in 2010. But the bird's range hardly increased at all – there were more of them in Devon but they hadn't made the break to Somerset, Dorset or even to Cornwall.

For a species which had been so widespread so relatively recently, and which was now so narrowly restricted to one geographical area, again it seemed sensible to try to set up some new populations in new (but formerly occupied) sites in order not to have all one's cirl bunting eggs in the same south Devon basket.

But this reintroduction would be of a very different feather. How do you rear and release the large numbers of baby buntings that would be needed? Paignton Zoo became the RSPB's partner in this project and the outline plan was to take early chicks from early nests (which usually fail anyway), rear them in captivity and then release them in the autumn into a site in Cornwall.

After a feasibility study in Devon the project started in earnest in 2006. By 2010, more than 300 birds had been released in Cornwall and 16 pairs were breeding there. It seems that this project is edging closer to success. And again it has involved working with local farmers. Maybe a site in Sussex or Hampshire should be the next leap forward?

Great bustards

The RSPB was a bit sniffy about the proposed reintroduction of great bustards on Salisbury plain for quite a while. Without going into the details, I think that some of our concerns were more justified than others, but I took the opportunity of being in Bournemouth in 2007 for the Labour Party conference to call in on the Great Bustard Project on my way home. That meeting established a closer relationship with the Bustard Trust which in turn led to a successful bid for EU funding to put the project on a much more secure financial footing. And two years later in May 2009 the project was able to announce the first successful nesting of the species in Britain since 1832 – the year of the Great Reform Act when the middle classes (only men of course) got the chance to vote.

The project is going well now and it seems more likely than not that a small but viable population of great bustards will become established on Salisbury Plain. That's quite an achievement. Who knows – with a bit more recreation of chalk grassland habitat, there might be quite a few great bustards in the heart of southern England in a few years time?

Cranes

It's a long time since cranes nested commonly in the British countryside – something that large which nests on the ground isn't likely to survive in a land of bows and arrows let alone one with firearms. I saw my first crane in Norway on the Cambridge expedition of 1978 on a morning when my tent had collapsed due to the weight of falling snow overnight. I decided I'd get warmer by going for a walk than by trying to fix the tent and wandered off only to spot a crane flying into a large shallow wetland valley – a quite different-looking place from most of southern England! But many cranes nest in eastern Europe in farmland landscapes little different from large wetland areas such as the Norfolk Broads or the Somerset Levels.

A small crane population did establish in the Norfolk Broads of its own accord but our judgement was that it would take a very long time before it could expand into other suitable areas. A great deal of wetland restoration has been done over the last few decades (and much more needs to be done) but the cranes seemed a little slow on the uptake. Could we give them a hand?

A collaborative project between the RSPB, the Wildfowl and Wetlands Trust and Pensthorpe Conservation Trust decided that the Somerset Levels would be a place where cranes would do well and which they would be unlikely to colonise for many years left to their own devices. The project now brings crane eggs to Slimbridge from eastern Europe, where they are hatched and the cranes hand reared by staff dressed in crane costumes so that the chicks do not grow up thinking that they are really people. The young birds are then moved to the Somerset Levels where they are kept and fed in a pen at a secret location. When they are considered ready, they are allowed to come and go as they wish and, so far, the released cranes are doing very well and making tours of the Levels. The first batch of cranes was released in 2010 and I was able to visit them, don a crane costume (as though that would fool anyone!) and have a look at the birds while they had a look at me too – pecking at my Wellington boots.

The 18 first-year birds were joined by another 15 released in the summer of 2011 and the Great Crane Project website shows that the Aller Moor area is the best bet if you want to try to see a Somerset crane.

White-tailed eagles

Most people seem to have come to terms with red kites and even grown to love them – despite the widespread fear and loathing of raptors by some 'real', 'country' people.

Nobody outside the conservation community is very worried one way or the other by cirl buntings, and cranes are generally popular, but a proposal to bring a large eagle back to southern England stirred up a hornets' nest of dissent which helped to knock the idea on the head – or at least to delay its progress.

White-tailed eagles lived in Scotland until the last one was shot on Shetland in 1918. But again we shouldn't think that remote coastal Scotland is the only type of place where white-tailed eagles are able to make a living, catching fish and feeding on seabirds and carrion. The Scottish population, which of course is derived from an earlier introduction (some of which I saw whilst following red deer around on Rhum in 1977), lives in much the same environment as birds in Scandinavia, but many of Europe's white-tailed eagles live in countries like Poland, and Belarus where they inhabit flat wetland areas with inland freshwater lakes – a bit like East Anglia.

I regret that we didn't get this project to the stage of releasing birds and some of the reasons for that were our fault and some were definitely not. The project was led by Natural England, with the RSPB as a close partner, but as the 2010 general election got closer and closer the clearer it became that Natural England were going to have their budgets slashed by the incoming Conservative Government (actually in cahoots with the Liberal Democrats as it turned out) and that large birds of prey didn't suit the incoming Conservative Ministers. That was the nail in the project's coffin but there had been a couple of other hiccups too. We in the RSPB had a wobble over bitterns which, in retrospect, was a bit silly since bitterns and white-tailed eagles live together over much of the eagle's range (and bitterns are doing spectacularly well now). And, in addition, pressure was put on an interested Norfolk landowner not to cooperate with the release plans.

Despite the fact that businesses in the local tourist industry were very keen on the proposed reintroduction to coastal Suffolk, the local Country Land and Business Association trotted out a lot of scare stories about eagles scaring and eating free-range pigs and poultry – rather to the embarrassment of their HQ colleagues, I always thought. Placards were posted along major roads appealing against the reintroduction plan, and the press loved the conflict.

When Natural England pulled out their funding the RSPB had to decide whether to go it alone, and this was at a time when we were nervous about our own future funding and knew we would face fierce opposition from a few landowners. So we shelved, but never completely abandoned, the idea.

I still think that a white-tailed eagle reintroduction would be a great thing to do in East Anglia and if there is a group of people out there who want to give it a go then count me in. I wonder how much money could be raised simply through appealing to the public – a Big Society solution to the funding problem.

What next and what not?

I am an enthusiast for reintroduction projects but only for those species that really need them. If a species will make it back to its former range fairly quickly on its own then let it get on with it. And if the practical details of a reintroduction seem insurmountable then let's leave it for later – but let's not be put off just because things are difficult. Most good things involve a bit of sweat and worry and RSPB staff almost always rise to a challenge.

There were a couple of potential reintroduction projects which came up in my time at the RSPB which I think we called correctly – chough and osprey, both in England. In both cases the birds found their own way back to England, the choughs probably from Brittany (which was a bit of a surprise) and the ospreys in the foot-and-mouth year of 2001 to the Lake District. Interestingly, both are now tourist attractions and are making their own sizeable contributions to the economies of their adopted communities.

So which should be the next bird reintroduction projects? The white-tailed eagle can be regarded as unfinished business and I hope that one day someone does get this one off the ground. But I don't see many other exciting species with strong credentials for reintroduction. Species lost relatively recently, such as Kentish plover, black tern and red-backed shrike, might be given a boost by climate change and not need any active help. Long-gone species, such as Dalmatian pelican seem to me to be difficult to justify. To my mind, reintroductions in future are likely to be useful in two situations. The first is local reintroductions of existing UK species where severe population declines have removed them from particular areas. Maybe future nature conservationists will be reintroducing corn buntings to the east Midlands or tree sparrows to Kent rather than working on national reintroduction projects. The second case will be in giving a variety of less mobile species a helping hand as climate change creates the conditions in which they may thrive in places they cannot reach. We may need to learn how to leap-frog species to new suitable habitat patches.

Introductions are different

Conservationists put a massive amount of meaning into the letters 'r' and 'e' when they compare introductions and reintroductions. Reintroductions are generally seen as a good thing because they put things back where they once were, giving nature a helping hand, whereas introductions put things where they have never been naturally. The difference is very significant because introductions are one of the major causes of species extinctions across the world.

Introduced species have a pretty shabby record in the UK. Grey squirrels are lovely little critters, and you only have to see a child feeding a squirrel in a London park to see how that interaction could fire a life-long interest in nature. But they are responsible for the demise of our native red squirrel. Red squirrels have been pushed to the northern and western reaches of the country by competition with grey squirrels and by the diseases that grey squirrels have brought. Muntjac deer are eating their way through our woodlands and gardens after escaping and being deliberately released in the early 20th century, while they should be looking cute in the forests of India or China. And mink, escaped or released from fur farms, are thought to be one of the primary reasons for the decline of ratty, our native water vole.

There are certainly a range of introduced bird species (and many more plants and mammals) that may cause bigger problems in the future. Will ring-necked parakeets (from India) become pests of fruit farms and maize crops, or become a serious competitor with green woodpeckers or barn owls at some stage? Will the spreading Egyptian geese (from Africa) interfere with native wintering wildfowl, or perhaps other waterbirds?

It is so difficult to tell which introductions might become a problem and which might not. Just because some introductions cause havoc doesn't mean that they will all do so. And so the guiding principle must be to prevent the risk of introductions happening in the first place. The craze for global free trade doesn't help – it is now easier than ever for species to be moved accidentally, and sometimes deliberately, around the globe. Many plants, insects, fungal diseases and a few birds are bound to be added to our native fauna and flora over the next decade or so, though thankfully the banning of the wild bird trade has limited some of the risks of new avian introductions.

And there are two species, both owls, which usefully illustrate the fact that these issues are rarely black and white.

The little owl takes us back to Lilford Hall as it was the same Baron Lilford who wrote about red kites who successfully introduced the little owl into the UK. Until then the range of this small owl had stopped at the English Channel. It was on St George's Day 1889 when Lilford's gamekeeper found a little owl on a nest. As Baron Lilford wrote:

> *I have for a considerable number of years annually purchased a number of these Owls in the London market, and as the majority are too young to fly when first received, I have had them placed*

> *together in large box-cages in quiet places about our grounds in Lilford, and left the doors of the cage wide open, taking care that an ample supply of food is provided once during the day for the Owls. I regret that I have not kept notes as to when I first adopted this practice, but for several years, beyond the fact that we occasionally saw and frequently heard of one or more Little Owls in the neighbourhood, nothing of importance came of the experiment, so far as I am aware, till 1889, when one of our gamekeepers, on April 23, found a Little Owl sitting in a hollow bough of an old ash tree in the deer-park at Lilford; she would not move, but he lifted her, and found that she was sitting upon a single egg, to which she added three, and brought off four young birds in the second week of June. One, if not two, other broods were reared in our near neighbourhood in 1889.*

His Lordship invested his time, money and enthusiasm into getting this owl established but I've always wondered what the RSPB would say if he came to them now and suggested the same scheme. He might have been given short shrift as the little owl is a non-native species – even though it is a close neighbour and lives happily on the far side of the English Channel. We don't suspect the little owl of any great harmful ecological impact on our farmland wildlife so it appears to be one of those introductions that is benign – and that is to be expected given its near-neighbour status which is rather different from the provenance of those troublesome introductions from the USA (e.g. ruddy duck) or Asia (e.g. Muntjac deer). In fact I wonder whether there are any harmful introduced species of European origin?

Which brings us to the species where all these considerations get mixed up together – the eagle owl. It's a cracking bird and I have enjoyed seeing them in the hills of the Alpilles north and east of Arles in the south of France. And although there are fossil records of eagle owls in the UK from about 7,000 years ago there are no certain subsequent records – although at around the time of writing one has apparently been seen flying in off the sea in Norfolk as part of an unprecedented influx of short-eared owls.

The tricky issue with eagle owls is that they sometimes turn up in upland areas where they may occasionally harry and sometimes kill hen harriers. An eagle owl was filmed in the Forest of Bowland in 2010 apparently killing a female hen harrier which made it a rather unpopular bird. The question is – or is it? – is the eagle owl a native or a non-native species? There is little doubt that all, or maybe just nearly all, of the eagle owls wandering around the UK are escapes or releases from captivity. But if this is a native, even if long-ago-native, species then does it matter whether the current crop of eagle owls come from cages or fly across the English Channel? I think that because it is a long-gone native species *and* a near-neighbour on the continent we just have to accept the occasional depredations of eagle owls on other species of conservation concern, in the same way that we would the impacts of carrion crows or foxes – that is be fairly tolerant but prepared to intervene (not necessarily with lethal intent) if the conservation impacts become too extreme. But the little owl and eagle owl show the grey area between native and non-native species and the need to think through the difficult cases and come to sensible positions even if they are not always universally popular. There is no reason why nature ought to be simple to understand or simple to assess.

Conclusions

Reintroduction projects form a small but quite useful weapon in the nature conservation armoury. This option should not be used too often but nor should it be shunned when it can give nature a big leap forward. We'll have to wait and see whether the projects underway at the moment end up as successes or failures but I think that they have all been worth a throw of the dice.

On the other hand, deliberate or accidental introductions are potent threats to our native wildlife and more should be done to limit the chances of them occurring and to minimise their impacts.

> *The wise man should restrain his senses like the crane and accomplish his purpose with due knowledge of his place, time and ability*
> **Chanakya**

Chapter 9

Nature reserves

Everybody needs beauty as well as bread, places to play in and pray in, where nature may heal and give strength to body and soul
John Muir

As you may remember from Chapter 1, I was a volunteer at RSPB nature reserves at an early age and when the RSPB *BIRDS* magazine arrived I always turned first to the recent sightings and news from the reserves. For many of us the RSPB is mostly about places – places with birds and a rich range of other wildlife – places like Minsmere, Loch Garten, South Stack and Belfast Harbour. And in those four reserves alone, picked to represent the four UK countries, you could see a high proportion of all the birds regularly occurring in Britain if you visited them all on a regular basis.

Nature reserves are places where nature is given the upper hand in decision making, rather than having to pick up the crumbs after people have had their fill – and the whole business of acquiring and managing land requires a good deal of work.

Buying up Britain – why it's worth it

The UK occupies about a quarter of a million square kilometres and about 60 million of us live there. So, on average, about 250 people squeeze onto every square kilometre of British land, so that your share of the country, if it were divided out equally, would be about 0.4 hectares. That's quite a thought.

It was Mark Twain, an American, who coined the phrase 'Buy land, I hear they aren't making it any more' in the late 19th century, when the population density of the USA was only 8 people per square kilometre (it is now about 32 people per square kilometre). The UK has the 53rd highest population density in the world: higher than Germany (56th), China (80th), France (98th) and the USA (179th) – but less than an awful lot of small city or island states, as well as the Netherlands (30th), India (33rd) and Belgium (36th).

Land is indeed a precious commodity and in a world of growing populations it will become more and more precious for growing food, fibre and providing housing. And that will put the nature which depends on that land under increasing pressure. So it makes sense for nature conservation organisations to get their hands on some of the best bits, as well as trying to influence the use of land on a large scale, even though the amount of land they can own will only ever be rather small.

After more than 80 years in the business, the RSPB's land holdings amount to over 143,000 ha. That's an area a bit bigger than Greater Manchester and a bit smaller than Greater London. And yet, practically every bird species breeding in the UK nests on one RSPB reserve or another. On those same nature reserves you could see, if you were very well-informed, diligent and lucky, a quarter of all the plant and animal species in the UK – on just over 0.5% of the UK land area.

These figures highlight the fact that wildlife tends to be clumped – some places have lots of it and some have rather little. If nature reserves are to be a powerful conservation tool then you need to make sure to get your hands on the land that has most nature, so as to make the biggest difference in terms of protecting it from future threats. Or you have to get your hands on those places where you can make the biggest difference and turn a wildlife-poor site into a wildlife-rich one.

Land ownership gives you power, opportunities and rights. That's why land isn't free – you have to buy it! If somebody else buys land of high nature conservation value then that value may well be lost over time, but if a nature conservation organisation buys that land then nature should, and usually does, thrive on it.

Getting out the cheque book

My boss for most of my time at the RSPB was Graham (now Sir Graham) Wynne whose background was in town planning but who joined the RSPB as Reserves Director at around the time I joined the staff as a humble biologist. Graham was keen on land, and under his leadership the RSPB increased its holdings considerably. When Graham became Chief Executive and I followed him as Conservation Director (which by then incorporated the strategic reserves function) there was no diminution in our shared enthusiasm for spending the Society's money on safeguarding sites for wildlife. But land does cost serious money and we didn't have free rein.

The RSPB spends a variable amount of money on land purchase every year depending on whether it is feeling skint or reasonably well-off, what land comes available, and what other monies can be found to add to the Society's own resources. But in an average year the Society will spend £1–2m on acquiring land, although there was one heady year when we spent about £10m.

The average UK house price in late summer 2011 was just under £0.25m so spending a net, say, £1.5m on land purchase is only like buying half a dozen houses – it's not a very big deal for an organisation with a turnover of getting on for £100m per annum. Buying land is, actually, a bit like buying houses and brings its own pitfalls, disappointments and legal work. But it gives long-lasting joy when you get it right, even though you do then have to knuckle down to the business of meeting the costs of upkeep.

Like buying a house, you are in the hands of the vendor to a large extent. The RSPB has an excellent Land Agency team, now led by Ian Baker and formerly by Geoffrey Osborne, which constantly has its feelers out for new land deals. It is in contact with lots of landowners adjacent to and near existing RSPB nature reserves, as well as with those who own land in areas where the RSPB would like to establish a land holding. Again, like buying a house there is an almost unlimited amount of land for you to buy each year if you want to – but what's available is not necessarily where, or what you would want it to be, or at the price that you were hoping for.

Sitting at the RSPB Headquarters in Sandy, one was bombarded with ideas for land purchases from colleagues across the UK. Each RSPB regional team would think they had found the perfect bit of land for RSPB purchase and there can be no clearer case for the need for an HQ than the need for someone to make those difficult decisions, balancing the interests of nature conservation and the RSPB. So, you must find ways of saying 'no' to some, indeed most, of your colleagues' land-purchase suggestions, and a way of negotiating a corporate 'yes' for the chosen few.

I won't say much about the former process as that has to do with the internal workings and politics of my former employer, but for the rest I relied a great deal on the common sense, political understanding, and sometimes downright stubbornness of first Gareth Thomas, and then Gwyn Williams, with an occasional serving of heavyweight help from me.

You can't say yes to everything without spending the budget ten times over – but you do want to say yes to as many propositions as you can. And you have to factor in that, like house purchases, some deals just fall by the wayside. So you must follow potential purchases that exceed your budget because you know that some will fail, but not which ones.

The RSPB is perceived as rich by many landowners, who neglect to notice that it is a charity with financial reserves that would keep the organisation running for barely ten weeks, if the money stopped coming in. Whereas a private buyer can spend what they like on a land purchase, within their means, a charity must be able to justify its expenditure and so can't and won't pay 'silly money' for land. Many prospective land deals have foundered on the unrealistic expectations of the 'vendor' who ended up not selling anything. My advice to anyone wanting to sell land to an NGO, for I am sure that most are just like the RSPB, is to look for a fair price near the district valuer's estimate, and not expect to make a killing. It'll be a private purchaser who pays a ridiculous price for your land not a charity with experience of land purchase and lots of potential purchases all over the country where they can spend their money wisely.

Once a land purchase looks like a good bet to staff then Council approval is needed – quite rightly. The RSPB Council meets four times a year and new nature reserves, or significant extensions to existing ones, need Council approval, which means covering all the bases and providing Council members with the case for purchase including all the snags and how they will be dealt with. Council usually sits on the enthusiastic side of the 'will we, won't we?' debate but, just occasionally, and usually rightly, are sceptical enough to require staff to rethink.

Lining up potential funding for a land purchase has to be done in advance of being sure that it will go ahead – and the process is fraught with difficulties. An appeal to support a land purchase is occasionally made to the RSPB membership and this has to be factored into the process too. People being people, the last few days of the financial year were always very busy for our land agents and solicitors and I was kept busy signing lots of bits of paper as part of the legal processes involved. Often millions of pounds would finally be spent on the last day or so of the financial year and then the whole process would start again for next year.

Nature reserves as teenagers

I tend to think of nature reserves as teenagers – many of them are absolutely lovely but they'll be even better when they grow up a bit. It's very rare that exactly the perfect plot of land comes onto the market just as you have the money to purchase it straight away. Land usually becomes available in inconvenient plots, at inconvenient times. Making a nature reserve is like putting a jigsaw together – but you can only buy different parts of the jigsaw at different times. Like buying a house room by room – you may end up living in just the spare bedroom and the bathroom while you wait for the kitchen and living room to come onto the market!

Graham Wynne used to cite the long history of Minsmere as an example of this process. Since arriving at Minsmere in the late 1940s, the RSPB has extended the reserve bit by bit over the years when money was available and when land came up for sale. Now Minsmere's boundaries stretch, not completely unbroken, south across grazing fields towards Sizewell nuclear power station and north through Westleton Heath almost to Walberswick. That makes for an exciting and varied nature reserve but it wasn't in any sense a one-off purchase. While the first parcel of land bought at Minsmere actually made complete sense on its own, sometimes you have to creep up on a nature reserve over a long period of time.

Protect or provide?

Nature reserves do many things – some make money for the organisations that run them (but most do not), and some are great places for recruiting members or educating children. But all of them have a role to play in maintaining and enhancing wildlife. Some nature reserves were acquired to protect existing sites of high nature conservation interest and others started life as wildlife-free areas which had massive potential. The dichotomy between protecting areas of existing interest and creating new wildlife-rich ones is a powerful mix.

In an earlier chapter I discussed the statutory protection given to important wildlife sites and you might think that if the best sites are protected in this way, as they generally are, then it doesn't matter who owns them. There is some truth in this, although the legislation is better at stopping bad things from happening at a site than it is in fostering good things. And you can never be sure that a bad or ignorant future government might not remove, downgrade or ignore the legislation that protects even the best sites. So it makes perfect sense for organisations like the RSPB to own sites such as seabird colonies, parts of internationally important estuaries or areas of heathland. And that's why it is such a good thing that organisations such as the RSPB, Wildlife

Trusts, Wildfowl and Wetlands Trust, National Trust, National Trust for Scotland and others actually have many important sites in their ownership, under their care and in their hands.

I wish that nature conservation organisations owned many more of the very best wildlife sites in the UK.

But over time nature conservationists have tended to shift their attention to the other end of the scale – recreating wildlife-rich areas from scratch! Habitat recreation puts nature back and is a very important aspect of the modern nature conservation scene. But no-one would claim that it is cheap.

Lakenheath Fen is a very successful example of habitat recreation. Located on peaty Fenland soils and set alongside the Peterborough-Norwich railway line, I often used to look out at Lakenheath from the train on my way to a meeting with the BTO at Thetford. It was easy to tell when you had reached the RSPB land as it was completely different from that which dominated the landscape for miles around. Lakenheath is a wetland with reedbeds, wet grassland and a few small patches of poplar trees. Walking through the reserve on a day in May you may hear golden orioles singing from the poplars and bitterns booming from the reedbed – and if you are lucky you might see cranes walking across the grassland. The sedge and reed warblers will be singing their hearts out, marsh harriers will cruise the reed-tops, bearded tits will be 'ping!'-ing, and flocks of hobbies will be chasing dragonflies. Judging from its birds this is a vibrant wetland and if you knew no better then you might just thank the RSPB for protecting this site from the drainage and intensive agriculture that dominates the surrounding areas.

But if you thanked the RSPB for that, then you would actually be mistaken and should really be giving even greater thanks – only some 15 years ago, Lakenheath Fen was a patch of carrot fields with no more wildlife than its surroundings. This is an example of habitat recreation *par excellence* and has led the habitat recreation charge in the Fenland. It shows how quickly wetland habitats can be re-engineered and to what spectacular effect.

And re-engineered is quite a good way of putting it because it is a bit more complicated than 'just add water, stand back and enjoy the wildlife'– although, of course, water is the key ingredient. There has to be enough good quality water and it must be sufficiently well controlled that you aren't going to annoy your neighbours by leaving them short or providing them with too much (i.e. flooding them out!). When what is now a wetland wildlife oasis was still a bunch of carrot fields the RSPB recognised the potential for wetland recreation here and set about the land purchase with enthusiasm. This initiative was taken soon after the death of the RSPB's Director General Ian Prestt in 1995 and Lakenheath was set up as a memorial nature reserve to recognise his great contribution to nature conservation.

The aim with Lakenheath was to recreate a wetland for bitterns and a range of other wetland species. Bitterns were certainly a target species as their population was very low at the time and very vulnerable to habitat loss caused by rising sea levels, most of their breeding sites being in coastal reedbeds in East Anglia. So the aim was absolutely to create a wetland where bitterns could thrive away from those threats.

These future flocks of bitterns that would darken the East Anglian skies were a cause of concern to the local RAF base at Lakenheath, which is largely populated by the US Air Force. The RSPB's plans were delayed while military authorities pondered the threat to our defence capabilities posed by a few brown herons with a booming song and we heard that our plans were discussed at length in the Pentagon of all places.

Once the military were comfortable (-ish!) with our plans there were further administrative hurdles to overcome, this time relating to the Reservoir Act. Although we didn't think we were creating a reservoir apparently we were! While all that was sorted out, and before the real work could begin, we made a decent amount of money out of carrots for a year or two. Finally, the diggers and bulldozers could move in under the direction of Norman Sills and land shaping and forming was the name of the game. Thousands of reeds were planted – many by groups of volunteers working by hand. Then we did reach the stage of 'just add water' – and could stand back, let nature take its course and see whether our hopes would be realised.

The reed buntings, sedge warblers and reed warblers that inhabited reedy ditches in the surrounding Fenland fields flourished at Lakenheath – their numbers increased a hundred-fold and their songs filled the spring air. Marsh harriers came back quickly and bearded tits followed. Bitterns began to be seen in winter and then grunting males stayed into the spring. Eventually it was clear that bitterns were nesting and their feeding flights in summer confirmed that they had chicks in the nest. In 2010 there were five bittern nests in the former carrot fields and two pairs of cranes have nested for several years.

Do visit Lakenheath and enjoy a spring walk there. You may see water voles and dragonflies and a range of wetland plants as well as the burgeoning birdlife. It's a lot more than 'just' a site for bitterns. It is an exciting recreated wetland that contributes to the protection of UK wetland wildlife in a landscape dominated by highly intensive agriculture. But the bitterns are perhaps its greatest success story – the five nests at Lakenheath now represent some five percent of the current UK population but back in 1995, when it was just a bunch of carrot fields, those five nests would have been closer to 50% of the UK total.

The RSPB is sometimes criticised for its species-oriented approach to nature conservation but I think that Lakenheath demonstrates how unreasonable that criticism is. I took a senior civil servant to Lakenheath once and it opened his eyes to what habitat recreation and targeted nature conservation action actually meant. I think he may have been told by others that Lakenheath was a bittern zoo or some such nonsense, but a walk through the reedbeds demonstrated to him the wealth of life that now thrived on this plot of land that had been, only a decade ago, unremarkable in the greater farmed landscape. The fact that the RSPB had bitterns foremost in mind whilst setting up this wetland was important, and that the bitterns had responded, was much more than the icing on the cake. But even if the bitterns had shunned the place (and why should they?) Lakenheath would still be a success because of its contribution to the local community, its otters, its carbon storage and its cranes and other birds. All that and the US Air Force can still fly in safety!

Habitat recreation projects like Lakenheath are expensive as, ironically for a nature-lover, the commercial value of these wildlife-rich habitats is low compared with the farmland on which they have to be created. Carrots are worth more money than cranes, apparently, and so if you want a crane-field you have to pay the current carrot-grower a fair sum to see off all the other carrot-growers who will be bidding for it. If Lakenheath had survived as a fen through recent history then it would have had every designation that a grateful nation could bestow on it – and its land value would be low because its economic potential was low. But because it had been stripped of its wildlife centuries ago then the RSPB had to spend a small fortune to put some of the wildlife back. It's a funny world.

Nature reserves based on habitat recreation schemes have become very much more numerous over the past couple of decades. The emphasis has shifted somewhat from protecting the existing wildlife-rich hot-spots to recreating some of those that have been lost. In The Fens, the National Trust, the Wildfowl and Wetlands Trust and the Wildlife Trusts all have their plans for doing something similar in many respects to the RSPB initiative at Lakenheath. We may see a small but important renaissance of wetland wildlife in this part of the world.

But around the country there are many other projects ongoing. Many of the habitat recreation schemes focus on wetlands but chalk grasslands, coastal saltmarshes, heathlands and some types of woodland are also being recreated. Much of this work is novel and imaginative, much is as yet unproven, but it is the way to go, and 'doing is learning' when it comes to this type of thing.

Management is key

Once you have bought your nature reserve the hard work really starts. There are hardly any sites where you can just do nothing. The closest we get to that in the UK, I guess, are seabird cliffs where most of what you need to do is to try to stop people falling off the tops of them while you let the birds get on with it.

In the UK, we live in a countryside where the print of the human hand lies heavy indeed. In its pristine past there were no fields, no conifer plantations, or railways and canals. If nature were given a few hundred years left to its own devices, much of southern Britain would revert to woodland and our rivers would reclaim their valleys. But since we aren't leaving nature to get on with it, and since we are still trying to make a living from the land too, then those small areas which are given over to nature reserves are, to a large extent, dependent on ongoing management to retain their interest and nature conservation importance.

Visitor infrastructure such as signs, paths and hides require upkeep but so do reedbeds, heaths, grasslands and woods. And once you have decided to intervene then you need a plan for what to do, and how and when to do it. The RSPB and some other nature conservation organisations produce management plans for their nature reserves which specify what management should be done and why. These are updated every few years and form the basis for management decisions at over 200 nature reserves across the UK – ranging from urban fringe sites such as Saltholme on Teesside and Rainham Marshes inside the M25 in Essex, to moorlands on Yell, Shetland or Geltsdale in the north of England.

And management costs money. And because you want that money to be spent in the best possible way for nature conservation, the management has to be managed. Management is expensive and resources have to be planned and allocated. Compared with the costs of buying land initially, the running costs of habitat management are high, and of course, the commitment is continuing, not just for one year. As an organisation's nature reserve portfolio grows, long term planning of management costs becomes critical.

For wetlands, water management is key and that means having control over both water quality and quantity. Sometimes this requires storage facilities – which might look like reedbeds to you and me – so that wet grassland can be kept wet even through the dry summer months (although they aren't always needed in British summers!). In establishing a new wetland, a full initial survey and map is vital. Time after time I have stood in a field which looks as though it most certainly slopes in one direction, to be told, and always rightly, that actually it slopes in a quite different direction. Left to chance or to an inexperienced eye, like mine, a wetland creation project could go very badly wrong.

Grazing is a perennial difficulty – do you graze at all and if so how hard and how often? And with which livestock? And do you buy them yourself or let the grazing to others who own livestock? And then do you pay them or do they pay you? And whichever it is, what is a fair price?

We increasingly understand that woodlands need managing too. Although ancient woodland is often rich in wildlife its ancient nature does not mean that it has been left untouched for ever. Clearing, harvesting, coppicing and grazing have been part and parcel of woodland management for centuries and that type of management is still needed – at the right time and scale – in our woods today.

Then there are the people. Do you give them access to the entire nature reserve or try to keep some areas as sanctuaries, free from disturbance? And do people need boxes called hides to sit in to enjoy the wildlife or would they be better off with an uncluttered viewing spot? And do you need paths and signs or do you trust people to find their own way safely around a site? And what provision do you make for those whose mobility or sight is impaired? And do you have to provide a toilet block? Or a shop or a café?

Then there is predator control (as touched on in Chapter 5). Do you do it or not? And what is the threshold for action? And how do you do it – with a rifle or snares or traps? Which species are targeted and what non-lethal methods might work as well or better – or perhaps cheaper? Who will do it – RSPB staff (who might be keen or reluctant depending on their own beliefs and values) or do you employ the local gamekeeper? And if using ammunition do you use lead or not? And what are your animal welfare standards and how do you make sure they are fully applied?

There are clearly lots of questions of the 'whether?', 'how?' and 'when?' variety and the answers need to be matched to the local circumstances and the RSPB's overall policies.

It is essential to monitor the success or failure of management actions to inform future decisions. If your breeding lapwings dip in numbers then you need to know whether you were or weren't carrying out fox and crow control, whether you were or weren't practising the right sort of grazing (whatever that might be for the site) and whether or not the water level was too high, too low or just right, before you can vary things to try to do a better job in future. And the same is true if your lapwings soar in numbers – was it something you did that was responsible, or some happy accident that should inform future management plans. And how long do you wait before you are sure that you need to change anything? All species vary in numbers due to the weather, chance events and perhaps due to things that happen to them when they are away from their home nature reserve (perhaps as far away as the Antarctic in the case of Arctic terns) – so that population changes might be nothing to do with your management at all. Some of these questions can be answered by study but sometimes the real world is sufficiently complicated that you have to use common sense – that most mis-named of senses. I had to rely on the uncommon wisdom of Gareth Thomas, Graham Hirons and Jo Gilbert for a lot of my time at the RSPB and their understanding of these issues was not common in any way, but actually very deep and profound.

But you can't always manage everything

Wildfowling

People sometimes don't understand that shooting rights are separate from many other ownership rights. What this means is that sometimes you can buy a piece of land but not own the shooting rights to it – either because the vendor won't sell them or because they don't actually own them. This can create a situation where you own the land and can do with it what you will, but that others can come in and shoot game on your nature reserve. Although unusual, this does happen now and again. You are faced with the choice of getting the land and hoping, at a later date, to get the shooting rights, or passing on the land. Such dilemmas can sometimes be solved by agreeing with a vendor that they will retain the shooting rights and can exercise them for their lifetime, or for a limited period, after which they pass to the RSPB. This is a good example of the ways of the world not having been designed with the simple life in mind.

Introduced species

Generally speaking, a nature reserve is a place that welcomes nature, but that being the case some less wanted nature may make an unwelcome arrival, whether it be signal crayfish or grey squirrels from North America, or *Crassula*, a virulent pondweed from New Zealand. Sometimes it feels that your nature reserve is under assault from somebody else's nature.

Perhaps a less emotional example than some others is the constant battle to keep heathlands free of trees. Sometimes the sapling trees are derived from nearby commercial conifer plantations upwind. Each year tree seeds blow onto your nature reserve and some of them germinate, and every now and again you have to spend money on heathland management to clear the trees before your heath turns into a woodlot.

If these unwanted trees that blew onto your nature reserve were toxic smoke then the polluter might pay for the costs of restoration, but because they are seeds there is no recompense.

Water

Water comes from upstream and so the quality of the water entering a nature reserve depends on the land use of the upstream part of the catchment. Seventeen percent of SSSIs in England (by area) are still in poor condition due to water pollution – much of it from agricultural run-off – and there is little that the SSSI and/or nature reserve owner and manager can do about this on their own land.

Sea levels

Many coastal nature reserves in south east England are threatened by sea level rise. Sites like Minsmere and Titchwell are in the part of the UK where sea level rise caused by the expansion of warming oceans is compounded by the fact that the UK is still moving after the last Ice Age – the southeast of England is sinking at the same time as the waters are rising. Clearly, owning the land doesn't give you any control over these changes and it does no good for the warden to join a flock of knot to command the waves to retreat – as Canute himself never did.

Coastal reedbeds, like those at Minsmere but also at Walberswick and Benacre further up the coast, are threatened with saltwater incursion and in time there may be a very big tidal influx. This will wreck these reedbeds for bitterns – which unlike their American relatives can't cope with saltwater reedbeds – and is one of the reasons why creating inland reedbeds such as those at Lakenheath, Otmoor and Ham Wall is such an important idea.

The Climate

You can't control the weather and owning land makes no difference. But future climates are an increasing consideration in nature reserve purchase and management plans. If you are buying a wetland site then you may need to think about what the water availability will be like in 50 years time.

Fish stocks

Many RSPB nature reserves are seabird colonies from the gannetry at Grassholm to the steep cliffs of Bempton, and from South Stack on Anglesey to Rathlin Island off the north coast of Northern Ireland. It's relatively easy to provide safe and secure nesting sites for thousands of birds such as puffins, kittiwakes and guillemots but protecting the nesting sites doesn't do the whole job because these birds feed far out at sea and their feeding grounds need to be protected too. A job beyond the RSPB's reach.

A bunch of teenagers – a dozen favourite RSPB nature reserves

The RSPB currently owns or manages over 200 nature reserves and I have visited only about half of them, but still have some favourites. These are just some of the reserves that mean a lot to me personally and so may be an idiosyncratic selection. But it's mine, and I like it! In choosing those to write about I had to discard a great many contenders, and I do feel a bit disloyal to them. But the ones I have chosen tend to be ones where I have strong personal memories either because of my work or because of my leisure visits.

The Lodge

As well as being the headquarters of the RSPB, The Lodge, Sandy, Beds. is a nature reserve which includes heathland, woodland, a meadow and some ponds. And although not every visitor is a birder, there are enough who are that you'd have to believe that few species of rare bird pass through it unnoticed.

When I first started work at The Lodge there were still some house sparrows around but they are now long gone, and the days when tree sparrows were a common sighting are the stuff of legend. I seem to remember seeing woodcock roding more often in the past, but maybe that was because I worked longer summer hours in the office then?

For the years that I was Head of Conservation Science I had an office with a good view over the front of the main Lodge building – looking north. A hawfinch once sat in the tree outside my window and on another occasion a lesser-spotted woodpecker flew past. Pied wagtails would often nest in Stable Block and I would see them collecting insects on the lawn. In the winter there might be a grey wagtail on the roof among the pied wagtails.

A welcome confirmation that summer was nigh would be the return of the spotted flycatcher each May. One of the latest returning of our summer migrants, I would keep an eye open for them through May and expect to see them around mid-month. My first year in that office they appeared on 18 May and the next year on 19 May. In 1994 they were early, 10 May, but the next two years their arrival slipped back to 22 and 21 May respectively. In my last two years there I didn't see them until 24 June and 10 June respectively. And those later first arrival dates seen from my window were an accurate harbinger of spotted flycatchers becoming much more difficult to see – they are now later and rarer than at The Lodge then – probably a consequence of their overall decline across the country. If you had kept records over that period what else would you have noticed, I wonder?

When I became Conservation Director I moved to the other side of the building and overlooked the small ornamental pool on the lawn outside, with its koi carp. This provided a few more species with occasional visits from fishing herons, kingfishers and even common terns. The odd migrant green sandpiper would also pop in for a moment or two.

I remember the birds I've missed through being away from Headquarters at the time, being too busy or not wanting to get my suit dirty! Missed birds include: two-barred crossbill, ring ouzel, nightjar, honey buzzard and osprey.

Birds that I have seen include hen harrier, hawfinch, waxwing, Dartford warbler, woodlark and firecrest. Maybe the most amazing bird was a dead, and tailless, yellow-billed cuckoo found by the gardener under a holly bush one December.

Looking back at John Gooders's *Where to Watch Birds*, his entry for The Lodge suggests a summer visit would provide nightjar, redstart, tree pipit, lesser-spotted woodpecker and woodcock. I've seen the last three at The Lodge, although my only lesser-spotted woodpecker was back in 1998, and tree pipits are usually seen across the road, off the nature reserve, in the field that was set-aside for many years. I would be amazed if there are more than a handful of people working at The Lodge now who have seen more than a couple of these species there in the last decade, such are the declines in our woodland species.

If you haven't been to The Lodge for years there are three changes you might notice before you get out of your car. The first is that the brick wall that runs alongside the road before you turn into the drive is now in good repair. That wall caused us a lot of grief as we didn't own it but everyone assumed that we did – and blamed us for its decrepit appearance. When we purchased the woodland behind the wall it became our responsibility and we spent a fortune getting it fixed. Soon after, I saw a solicitor in Sandy to sign some documents who declared that it was good to see that we had finally mended that wall. As you drive down the drive, for you probably will drive, you may notice that the edges of the road are far more overgrown with nettles than they used to be. This is because of all the fertilising the roadsides get from the increasing numbers of cars that come and go every day as 500+ people come to work. But the major changes will be behind the wall, in its smart new condition, and to the right of the drive, where instead of a thick dark conifer wood you will see open ground with the faintest appearance, so far, of heathland.

Looming over the whole area is the Sandy Heath radio transmitter and a ridge of greensand rock that was once an extensive heathland – some of which remains on the site. The aim is to restore more of it, so that heathland species such as woodlark, nightjar and Dartford warblers can use the entire site and expand their ranges as a stepping stone to heaths further north. All three species have turned up here in recent years so watch this space.

The Lodge was my place of work for more than two decades and at either end of the day, before most people arrived and after most had left, it was a delightful spot with its attractive buildings, lovely gardens and enveloping nature reserve. In the mornings I would rush into work to start the day and get going, but as I left in the evening, on summer evenings in particular, I would stand by the car, drink in the scene and listen to a moment or two of birdsong, for that is what it was all really about.

Otmoor

Charles Dodgson, better known as Lewis Carroll, is supposed to have stood on the hill, maybe near the attractive pub, The Abingdon Arms, in the village of Beckley, and looked down over the patchwork of fields below him in the natural bowl that is Otmoor and thought of his heroine Alice crossing an enormous chess board and meeting strange characters such as Tweedledum and Tweedledee.

Carroll wrote 50 years after Otmoor was enclosed and that patchwork of fields was produced. Before then it formed more of a natural flood plain for the River Ray to flood in winter. And that's what the River Ray still tries to do each year with varying success as over the years it has been embanked and an infrastructure of drains and ditches has been built.

In recent years, the Otmoor area has been in part a Ministry of Defence firing range, and in part a Wildlife Trust nature reserve, but much has been farmland devoted to wheat. Otmoor is not really suited to wheat production unless you take extraordinary steps to keep the land dry. It's a big task and succeeds sometimes but not always. The RSPB purchased some of Otmoor back in the late 1990s and started turning it back into a more natural flood-plain, but this can only partly succeed when you welcome water onto your land while keeping it from spilling over onto your neighbours' land.

The future of Otmoor must surely be as a more natural floodplain that delivers water storage and flood mitigation services to the communities downstream and nature conservation locally. That doesn't mean that the site can no longer be farmed – an extensive livestock system would seem very suitable – but the route to such a future looks very uncertain because of the complexities of land ownership, individual rights and the economics involved.

In the meantime, Otmoor holds a large proportion of the breeding waders in Oxfordshire, is a potential new breeding site for bitterns (they winter there already) and in some winters has a really impressive starling roost. In late summer you can see brown hairstreak butterflies too.

I wonder what Lewis Carroll would think of Otmoor now if he could look down on its fields some 140 years after they inspired his writing. What tale would he now be motivated to tell?

Fetlar

Fetlar is amongst the most northerly of RSPB reserves – an island as far north of the English border as London is south of it. It is reached by ferry and is one of the best places to see killer whales in UK – although whenever I have been I have got the 'You should have been here yesterday' refrain from the various wardens.

Fetlar was famous from 1967 into the 1970s for being the only UK nesting site for that Arctic species, the snowy owl. But now the island is famous for holding most of the UK's nesting red-necked phalaropes – those tiny wading birds, which spin on the water and pick off insects, and whose males incubate the eggs while the females look for new mates. It's also a good site for whimbrels which breed on the moorland near the pools used by the phalaropes and not far from where the snowy owls used to nest.

These are all northern species, just nesting in the very north of the UK while their main populations are further north in Scandinavia, the Faeroes and Iceland. They are species that we would expect to see, or rather, not see, disappear because of climate change. Just as Dartford warblers are spreading north so might these three species retreat north – one spreading as warmer weather favours it while the others retreat back to cooler climes. Under these circumstances you might be tempted to write off these local populations and just not bother with them.

I think that would be wrong-headed because we can't be completely sure what will happen. If we can manage the pools where phalaropes live, so that their attractive chicks can hide from predatory skuas and find lots of insects to eat, then let's do that and keep them hanging on for as many more years as possible. I wouldn't die in a ditch to save red-necked phalaropes in the UK, as they are an abundant species elsewhere in the world and our phalaropes really are the edge of a huge global population – but I wouldn't turn my back on them either for the sake of a few thousand pounds and a bit of management effort now and then. And if I can go back to see them then, maybe, eventually I will see killer whales too.

Arne

Arne was famous in my youth as one of the sites where Dartford warblers hung on after the savage winter of 1962/63. The English population was reduced to double figures of pairs and several of them were found at Arne. By the time I volunteered there in 1975 Dartford warblers were common and nightjars 'churred' over the heaths.

I remember wandering off across the heath one evening to listen for nightjars and just to enjoy the tranquillity of the place. As I walked along the path, looking through the gloom to make sure I didn't trip over a root or clod, I heard a loud shriek nearby. I fell over with the fright and shock of it as a Sika deer ran off across the heath and through the bog. I was glad it was dark enough and late enough that there was little chance that anyone else had seen me looking daft on my backside in the heather.

Sika deer are a non-native species introduced from Japan. Their numbers have become very high – enough to damage fragile heathland habitats – and on the advice of Natural England a programme of culling is now in place across a wide variety of land ownerships. Back in 1975 it felt like a Sika deer nearly killed me with fright but 30 years later I was involved in deciding that we needed to cull them at Arne.

It is a reserve to which I like to return. It's another site which is worth a visit at any time of year. The heath always looks good and the views across Poole Harbour or across Hartland Moor towards the ruin of Corfe Castle are striking.

You can see the rare sand lizard on the heath as well as adders, smooth snakes and slow worms. Grayling butterflies land on the dusty tracks and immediately close their wings and orientate their bodies so as to cast the smallest possible shadow.

On a visit in summer 2011 I saw the very rare ladybird spider being reintroduced to Arne. The males live for a couple of years and, if lucky, mate once and then die. The females however mate once, hole themselves up with their young, feed them as they die and are then eaten by their babies. All that might be going on next to your feet as you walk across the Arne heathland, so very rich in rare, threatened and fascinating invertebrates.

But I still think Dartford warbler when I visit Arne – I want to see a male perching on top of a flowering gorse bush and see this dark, long-tailed warbler dash between patches of gorse and heather. You can see them and hear them if you are patient and as you watch think about what a great conservation success story they are – they have increased more than a hundredfold in the last 50 years.

In contrast, over the last 200 years the area of lowland, heathland habitat has been reduced by something like 80%, which is why the protection and careful management of the remainder is of such importance and why we need the devoted maintenance of places like Arne.

Dinas

The conical wooded hill of the Dinas in mid-Wales is hidden away up the valley of the Afon Tywi. I first visited this nature reserve on family holidays as a boy and then once on a Field Club trip whilst at school. It's a place to go in spring when the migrant birds return from the other side of the Sahara, and wood warblers, redstarts, pied flycatchers and tree pipits seek out the oak woods of Wales.

My friend, the journalist Mike McCarthy of the Independent newspaper and the doyen of wildlife writers, and I visited the Dinas one wet day in May 2008. It so impressed him that Mike wrote about it in his excellent book *Say Goodbye to the Cuckoo*. But that visit was, for me, just the latest of many to this lovely spot.

After parking your car, you can take the boardwalk through the wet woodland by the riverside and stroll down to the Dinas hill itself. A circular walk will then take you around the hill and through the oak woodlands that clothe it. I always walk clockwise around this route – for some unknown reason it seems right that way around – and the path takes you first away from the river and slightly uphill through open woodland. Later the river is beside you and it's more difficult to listen for the sounds of birds but you should keep an eye open for common sandpipers and dippers along this stretch.

You'll probably soon see the red tail of a redstart hurrying into a nest box and the black and white male pied flycatchers aren't birds that hide themselves for long, but the bird I always want to hear and see is the wood warbler and that usually takes a little more time, even in this favoured haunt.

Wood warblers are in many ways 'just' little green-ish warblers, just like their close relatives the chiffchaff and willow warbler. But whereas I can easily hear and often see those other two species at my local birdwatching patch at Stanwick Lakes, I have to travel to experience a wood warbler. Once much more widespread their range is now predominantly western in the UK, with a strong preference for the Celtic fringe of Britain. But the Dinas is *the* place for wood warblers for me. I cannot think of them without thinking of the Dinas and its beauty, and I cannot visit the Dinas without thinking of wood warblers. The two are inextricably linked through long association and happy memories. Even if, as sometimes happens, I pop into this reserve in winter just to drink in its loveliness my mind will turn to the picture of a male wood warbler sitting on an oak branch in dappled spring light, his yellow chest producing the most glorious of songs to fill this most lovely of woodlands.

Wood warblers have declined more than many other species in recent decades and this may well be due to climatic changes here on their breeding grounds, or perhaps problems in their African wintering grounds far-away. Maybe because of these global changes they will one day be lost from the Dinas and even the thought of it saddens me as this is a bird of place for me.

Winterbourne Downs

There are some places that just should be nature reserves – and that's how I felt when a certain piece of Wiltshire farmland came onto the market. This farm, now the RSPB Winterbourne Downs nature reserve, would provide a bridge enabling wildlife to travel between the two great remaining chalk grasslands of Porton Down and Salisbury Plain, and is the type of place where it would be possible to restore the chalk grassland habitat for the benefit of butterflies and plants as well as a few birds.

So one Saturday in July 2005 I got off the train at nearby Grateley station and was taken around the potential purchase by Jane Brookhouse, the RSPB's Reserve Manager. It was a sunny day, which did nothing to dampen my enthusiasm, and there were dark green fritillary butterflies on the wing. I was already sold on the site from seeing it described on paper but an actual visit was useful to fix the picture of it in my mind, to understand the challenges of visitor access, and to see the buildings that would come with the land.

A few years later after the purchase was completed, the RSPB's Council visited the site to see the grassland restoration work being carried out there. The weather couldn't have been more different than on my first visit and the rain was torrential. We walked squelchingly around the site and later steamed in Newton Tony village hall. The contrast led me to wonder if my willingness to buy would have been as great if the weather had been that awful on my first visit!

Winterbourne Downs is not a highly promoted nature reserve as the access arrangements are not straightforward and the place is still very much work in progress. There is a small car park by a small meadow on the edge of the site and I have parked there now and again to look at the view and imagine how the place will look in future when it is fully restored. It has already been visited by itinerant great bustards so they are clearly keeping an eye on it too.

Nene Washes

The Nene Washes is not a very well known RSPB nature reserve but it is a gem. As well as most of the UK's nesting black-tailed godwits, and now cranes and reintroduced corncrakes, the reserve has one of the highest densities of breeding waders such as snipe, lapwing and redshank in southern England.

In winter, the Nene Washes, as with the nearby Ouse Washes, perform the function for which they were built by accommodating water that would otherwise flood the rich peaty arable land of The Fens. These two washlands were designed by Dutch engineers and built in the 17th century to provide areas into which the rivers could flood to keep the rest of the low lying Fens dry. The Nene Washes consist of two parallel earth banks, almost a mile apart which stretch about 13 miles through The Fens. In the winter months the fields are wet and filled with wigeon, wild swans, pintail, teal and shoveler, while hen harriers and short-eared owls hunt overhead for voles.

It is a nature reserve which is worth the visit at any time of year as there are always birds there. I've seen a Montagu's harrier fly down the Washes and once watched a barn owl which flew very close to me, looked me right in the eyes, screamed, and flew off.

Being close to Peterborough station, and the fast train back to London, I sometimes used the Nene Washes as a place to take people whom we wanted to impress. A variety of civil servants, politicians, donors and others have been picked up at Sandy station in the morning, taken to the RSPB's Hope Farm (see Chapter 7), brought on to the Nene Washes and then packed off back to London from Peterborough station.

The Nene Washes usually worked their magic on our visitors but a large part of that was due to the ability of the Site Manager Charlie Kitchin to answer any question that he was asked. Site Managers, or wardens, are all different and many are interesting characters. Charlie knows his birds very well (which is why I was delighted when one May morning I corrected his identification of a godwit flying up the Washes by pointing out that it was a bar-tailed, not a resident black-tailed) but he is also responsible for the management of the cattle that graze the grassland and so can answer agricultural questions with lots of knowledge too.

Like many nature reserves, the Nene Washes brings many issues together in one place – in this case predators and their control, wildfowling, reintroductions, grazing and farming, water quality and quantity, public access and disturbance, the right extent for the nature reserve, public perceptions and misconceptions about the RSPB, quite a few birds, and lots of other wildlife too. Whereas these issues might seem somewhat disparate to some staff at RSPB HQ, they all are intertwined in complex ways on the ground, and it's a guy like Charlie who has to live with them every day.

Some Site Managers are optimists and some are pessimists, and you have to know which you are dealing with when you visit and talk to each of them. I won't tell you which way Charlie inclines but will say that he was a great guide to his site for any visitor I brought there. I remember heading to the Nene Washes for the first time with a civil servant whose views were important to us, thinking that maybe I should have briefed Charlie on what to say and what not to say, and wishing I had got around to it. But I needn't have given it a thought, Charlie gave a very full account of the history of the nature reserve and an honest account of the challenges he faced there. After that I never even bothered thinking that I ought to brief Charlie – I just brought people to his reserve and let him chat away.

Different visitors want to know different things: a senior civil servant was fascinated by the intricacies of getting the Washes grazed, seasonally, in an arable area where cattle are thin on the ground; Peter Melchett of the Soil Association, and formerly of Greenpeace, was interested in the farming too but also let slip that he had been on RSPB Council when the Society decided to buy land here; and a local landowner was shocked that wildfowling was allowed on part of this nature reserve because the shooting rights were not owned by the RSPB. The Nene Washes work their magic on RSPB staff too. A group of colleagues who work in Scotland came down once and we met them off the train at Peterborough one May evening and ate fish and chips on the Nene Washes as the snipe drummed overhead and the godwits went 'wicka-wicka-wicka' in the wet fields nearby. Many of the RSPB's biggest and best nature reserves are in Scotland for sure, but my colleagues went away with a better appreciation that there were some jewels in the southern rim of the RSPB crown of nature reserves too.

Papa Westray

The RSPB has a whole clutch of nature reserves on both Orkney and Shetland. Most are primarily of seabird interest, as these northern isles between them hold high proportions of many UK seabird populations and in the case of gannets and great skuas a high proportion of the world populations too.

Maybe I shouldn't choose, but if I had to then I would have to say that I prefer Orkney to Shetland – it's a softer landscape with greater evidence of centuries of human habitation. Shetland is amazing – but amazingly harsh at times.

Papa Westray is a flat island that can be reached from the larger, neighbouring island of Westray, either by ferry or by the shortest regular scheduled flight in the world – it takes all of 90 seconds from lift off to touch down if the wind is in the right direction.

Papa Westray is home to what has been the UK's largest Arctic tern colony. Thousands of these birds nest on the ground on the short turf.

Arctic terns are very similar to, but really quite different from, their close relatives the common tern. Arctics have shorter, redder legs; shorter, redder bills; no black tip to the bill; longer tail streamers; greyer under parts; more uniform coloured wings; and a different call. But they are, despite all that, sometimes really quite difficult to tell apart. Tern workers sometimes suggest that the best way to tell the species apart is to wander into one of their colonies. If it is a common tern colony then the birds will dive down and peck you on the head whereas if they are Arctic terns they will do likewise – but draw blood. It's one way to attempt to tell them apart!

A vibrant Arctic tern colony is an amazing sight. In July, when they have chicks, birds come and go all the while, bringing individual fish one at a time to their ever-hungry offspring. Most of the prey are sandeels – long eel-like fish that do, as their name suggests, breed in sand banks and sandy seabeds. On Orkney and Shetland everything seems to eat sandeels – that's what a puffin has in its bill, and guillemots, kittiwakes and even the Arctic skuas, which make their living largely by stealing food from other birds, are feeding on sandeels. Out at sea, many fish eat sandeels too, and so do seals and porpoises. They really are a keystone species on which many other marine species depend.

But in the early 1980s, the populations of sandeels in these northern waters began to fall and increasing numbers of seabirds failed to breed successfully year on year. It takes quite a while to notice trends in seabirds because the adults are mostly long-lived and their breeding success does change quite dramatically from year to year – and different colonies sometimes have markedly different fortunes. But in Orkney and Shetland many species of seabird that depend on sandeels started to show very poor breeding success at around the same time.

Walking through the Papa Westray tern colony became an increasingly sad experience. Each year there were fewer birds nesting and in some years it was very obvious that chicks were starving to death. The changes in sandeel numbers may have been partly due to overfishing of stock locally or more widely in the North Sea – sandeel catches were even used to fuel power stations in Denmark at one time, although many are now used to feed farmed salmon.

Papa Westray's Arctic skuas also nest on its short turf. These birds are often called piratical because they make part of their living by stealing food from others birds – for example from Arctic terns. The skuas are dashing thieves – harrying gulls, terns or auks on the wing until the pursued bird drops the fish, usually a sandeel on Orkney, that they have expended their time, energy and skill in catching. For Arctic skuas, stealing is a way of life and they are very good at it, but it does mean that if your victims are finding life difficult then you may have a harder time too. If you are living off the cream of the land then if the cream turns sour you feel the pain.

This is also a place to see the rare Scottish primrose – a gorgeous little flower which grows in the short sheep-cropped turf. When the Arctic terns are sitting on their nests some of them will be looking at this tiny plant with its purple petals and yellow centre. Arctic terns spend our winters in the Antarctic and so get another summer – they probably see more sunshine in the year than any other species on Earth and cover huge distances in doing so. Contrast them with the Scottish primrose whose entire world distribution is restricted to a few sites in Orkney, Caithness and perhaps Sutherland. Both types of species need nature reserves – the globe-trotting and the sedentary – although the plight of the Arctic terns as their food supply dwindles shows that the nature reserve tool can only do part of the job for species like this, whereas it could provide a comprehensive solution for the range-restricted, endemic primrose.

Loch Gruinart

It's worth visiting Islay at any time of year – the whisky is always plentiful! But so are the birds, and if you are looking for birds then you'll always end up at the RSPB nature reserve at Loch Gruinart. In summer the Loch Gruinart flats hold lapwings, redshank, snipe and other waders as well as corncrakes. In winter the fields fill up with geese – barnacle geese and white-fronted geese from Greenland. At any time of year you may see a hen harrier slip past.

If you go in winter you will see thousands of barnacle geese. A little under half of the UK's wintering population come to Islay – that's about 35,000 birds – and you can see a fair chunk of them on the reserve at Loch Gruinart. They are from the Greenland population and the vast majority winter in north Scotland and Ireland. Not that far away, as the barnacle goose flies, lies the Solway Firth but the birds that winter there breed in distant Svalbard. The distance between the Solway and Islay is as close together as these two populations of the same species ever come – and there is equally little interchange between them and the Russian population which winters mostly in the Netherlands.

The history of barnacle geese on Islay has been chequered. They can cause damage to crops (although that damage is sometimes, to my mind, greatly exaggerated by farmers seeking compensation) and so the nature reserves on the island provide refuges for the geese and respite for farmers from goose visits to their fields.

And the draw of those birds throughout the year brings people to Islay who would not otherwise take the plane from Glasgow or the ferry from West Loch Tarbert. The wildlife is an economic asset that does wonders for the island's economy.

As an example, on a work visit one October we just had to pop in to the Bruichladdich distillery for a visit and now I am the part owner, with four other RSPB colleagues, of a cask of maturing malt whisky. So do go – and tell people you are visiting for the wildlife!

Titchwell

I first visited Titchwell as a schoolboy – before the RSPB bought the place. I was on my first holiday to East Anglia and I had spent a few days at Cley seeing my first ever bittern and bearded tits. The East Bank was the place to stroll according to my *Where to Watch Birds* book and looking over Arnold's Marsh that late July day I got talking to a man puffing on a cigarette and dressed in leathers who must have been the owner of the motorbike I had noticed parked by the road.

This man, I later learned, was the bird artist Richard Richardson and he showed me my first spotted redshank and talked to me about birds. Probably realising that this young teenager was a potential birder for life Richard told me about this place called Titchwell, down the road towards Hunstanton, where that year a pair of Montagu's harriers had bred and where you could now see the young harriers floating above the reedbed with their mother. It was very kind of him to let me know and after a bit of parent-pestering I was able to walk down the right track a few days later and see, over the reedbed, several orange-tinged young harriers in flight over the sandy-coloured reeds.

When visiting Titchwell these days, with its crowds, hides and visitor centre it is easy to be a bit nostalgic for those earlier days but that would be wrong. Richard Richardson was keen, I guess, to enthuse me about the birds he loved, and thought that letting me know where beautiful birds were would be a good way to do this, and I feel that there is a need for some (not all) nature reserves to be hot-spots for people as well as for wildlife. Managing the former so they can enjoy the latter without ill effect is just a task to do rather than a reason for excluding people entirely (except at a very small number of very sensitive sites).

I think of Titchwell more as a birdwatching site than a conservation site, which may not be entirely fair, but it is close enough to where I live to be on the list of places to a go at least a few times a year. And over the years Titchwell has provided me with pectoral sandpiper, grey phalarope, honey buzzard, Arctic redpoll and, of course, the late lamented 'Sammy' the black-winged stilt – even if there have been no more Montagu's harriers for me there.

And Titchwell is quite a good place to go to see the pressures on many of our bird species. Inland the farmland is intensively managed – it holds more birds than many areas but still shows only a shadow of its former wildlife richness. Out to sea you can now see wind turbines whose impacts on marine life have not been well studied or understood but which will at least generate energy far more sustainably than fossil fuels do. If you know the area well you will have noticed that the beach moves around quite a lot and the expectation is that it will move inland over the next few years because of climate-induced sea-level rise. Next to the nature reserve the saltmarsh is managed by a wildfowling club so all wigeon should learn to know where the boundary lies.

It's no surprise that birds flock to Titchwell and none at all that people do too. One of the most visited of all RSPB nature reserves, Titchwell generates considerable money for the local economy through all those visitors who use B&Bs, fill up with petrol, or just buy their Sunday papers on their visits.

The toasted sandwiches at the small canteen are a good buy, and as I munch on mine I think, sometimes, of a distant day when I saw my very first Montagu's harriers.

Rainham Marshes

A few minutes out of St Pancras, as you take the green route to Europe on the Eurostar, you pass the RSPB nature reserve of Rainham Marshes (on your right). As you speed past you can see wet fields and pools, a few reeds, the new visitor centre that always reminds me of a Bassett's Liquorice Allsort, and you might even be able to identify a bird or two. But the Eurostar won't stop and you'll have to take the half-hour journey from Fenchurch Street station to Purfleet if you really want to explore.

The RSPB acquired Rainham Marshes from the Ministry of Defence in 2000. It comprises a remarkable stretch of medieval grazing marshes, intersected by reedy ditches and silt lagoons, by the side of the River Thames, and all inside the M25. This rural landscape has been protected through a variety of mechanisms over centuries and was a military firing range for over a century. Graham Wynne once fired off a few rounds at Rainham in his schooldays and maybe that is one reason why he was always so devoted to the place.

Graham once took a potential funder of our work to Rainham in its early days. The site was still littered with junk that had been dumped there and it was certainly a bit rough around the edges. As they looked around a hoopoe appeared and perched on a burned-out car. Perhaps this was a sign that glorious nature will always return to a site given a chance, however rough it may appear to us.

Rainham was a great place for the RSPB to acquire – so good that I think we 'opened' it half a dozen times! Graham usually bagged these opportunities, which was fair enough since he had grown up nearby and put a lot of effort into the negotiations himself, but once I got the gig! There were no buildings on the site and so the event, for several hundred people, was held in marquees on the marsh. Luckily the weather was kind. I had to say a few words as did the RSPB's Vice President Bill Oddie. I don't know Bill that well but I've had to speak at events with him a few times and I've noticed that he always goes very quiet and withdrawn before speaking – he's clearly thinking about what he is going to say and running through it in his mind – but once he starts the impression is engagingly off the cuff as though he is just chatting.

I know I didn't see a water vole at Rainham that day because I've never seen one there. On many occasions a warden has almost promised me a sighting, usually saying that there were lots of them there the other day. But despite the fact that I know it is one of the best places to see this cute little mammal, and despite the fact that I have met many people who have seen water voles very well at Rainham, and despite seeing many photographs of water voles at Rainham, I have never, ever, seen one there.

You get a great view and a nice scone and cup of tea from the Visitor and Education Centre at Rainham – some would say that it is best to be inside the Centre because then you don't have to look at it, but I think that's unfair. It did win an architectural prize, which is often a pointer to a building in which one wouldn't want to live or work, and past which one might not want to drive, but to my eye it is a stylish erection and its modern look is appropriate to the site.

Look out from the centre and you can see the Thames with boats passing along it. Occasionally a train will speed past. Planes pass overhead all the time on their way to Gatwick, Heathrow, London City or Stansted Airports. There is a constant hum of road traffic and you can see wind turbines, a landfill site, factories, pylons and lorries. This is no rural idyll and yet it has wildlife at all times of the year. The only thing that Rainham has in common with, say, Aghatirourke nature reserve in Co Fermanagh is that they are both managed by the RSPB.

Rainham is a birder's site. I've seen spoonbill there and missed penduline tit, but the list of rarities is a long one and there's always something to see. And it's also a local's site. Many of the visitors get as much pleasure from the unusual sight of a grazing cow as from the birds – and some of them even see water voles, I am told.

Abernethy

Abernethy is a proper nature reserve – it's big. For a long time it was the largest RSPB reserve but now the Forsinard Flows reserve has overtaken it. But Abernethy is still big. They say that size isn't everything but at least when it comes to nature reserves it can be quite important.

Abernethy stretches from the top of the Cairngorms where dotterel play out their reversed-role breeding systems right down to the ancient Caledonian pinewoods where crested tits nest in ancient Scots pine trees.

This is landscape-scale nature conservation, alright. Abernethy is the largest single remaining chunk of ancient pinewood and the RSPB plans to double the size of that woodland in future. This reserve is home to rare and threatened invertebrates from ants to hoverflies, and of vertebrates from red squirrels to pine martens. As well as the famous pair of ospreys at Loch Garten, Abernethy provides a home for capercaillie, crested tits and confusing crossbills – some of which are of the UK's only endemic bird species, the Scottish crossbill.

The ancient forest of Caledon once dominated the north of Scotland just as ancient oak woods dominated the south of England. And similarly it was felled bit by bit over the centuries. The RSPB plans to restore this pinewood over a significant area. The complexities of the management measures required are informative. The main brake on the forest spreading up the hill, naturally and slowly but inexorably, is the grazing pressure from red deer. Red deer have no serious predators other than man in this wolf-less landscape, and their numbers have grown over the years to such an extent that they exert massive impacts on the ecology of the area.

Deer fences will exclude most deer from sensitive areas but the trouble with deer fences is that they kill capercaillie. Studies show that capercaillie are particularly prone to flying into deer fences placed within a forest, but as well as black grouse and red grouse they also die in collisions with the familiar wide mesh wire deer fences that cover much of the Highlands. So in Abernethy many deer fences have been removed, as they have on FC land and many Highland estates also. But the problem of deer control remains, so for many years deer have been culled at Abernethy to reduce their impacts on the forest.

But reducing deer numbers in some places seems to have allowed the vegetation understory to thrive, which would be a good thing if were not that in wet summers young capercaillie chicks may die if they get too badly soaked in wet vegetation. And wet summers are not exactly rare in the Highlands – and seem to be getting commoner with climate change. So it is possible that efforts to increase forest cover (good for capercaillie) through reducing deer grazing pressure, results in some areas becoming undergrazed (bad for capercaillie). Ouch!

I first visited Loch Garten as a teenager, keen to see the ospreys and delighted when I did. This was the site to which ospreys returned in 1954 after being driven to UK extinction in the UK for 38 years due to the legal, but still horrible, persecution of Victorian and subsequent egg collectors and trophy hunters. Osprey were also killed at one stage for their crest feathers and so became one of the species that the RSPB was founded to protect. Although the UK osprey population is now much bigger (over 300 pairs now nest in Scotland, England and Wales and they keep having a look around in Northern Ireland), it's still a joy to visit Loch Garten and see the birds through the mounted optics in the viewing centre.

In the summer that I graduated from Cambridge I used an interview at Aberdeen University as an excuse to hitchhike over to Abernethy and spend a couple of nights sleeping rough in the pinewoods. It's not a great way to get a good night's sleep and so I was pretty tired as I tramped around the woods listening to redstarts, watching crested tits go in and out of their nest holes, and seeing goldeneyes on the lochs. I dozed in the sun by the side of Loch Garten and woke to see an osprey lifting off from the water in front of me clasping a large fish – the bird's crash into the still waters had woken me up. A bit later I was wandering around an open piece of ground and an osprey was calling above my head. I realised I had come close to a nest which I could see in a lone pine tree a little distance away. For a brief moment I thought I had wandered underneath the famous Loch Garten eyrie and that I was being watched accidentally disturbing Scotland's most-loved pair of birds by scores of people, but it was a neighbouring pair. So I retreated in relief and spent the afternoon watching my pair of ospreys at their nest. To be honest, they didn't do very much, but it still felt like a privilege to share their company.

Abernethy would be worth a book on its own – someone should write one – but I have had a host of experiences in its ancient forest. I have pushed an aggressive male capercaillie away with my foot, persuaded myself that I had just seen and heard a Scottish crossbill only because the world expert on the species, Ron Summers, told me that I had, wondered why the forest-nesting greenshanks disappeared from these woods in the 1950s and pictured Desmond Nethersole-Thomson hunting them down, and watched roe deer and green hairstreaks, lekking black grouse and delightful red squirrels.

But there are two species that call me back to Abernethy. First, I would love to see pine martens even though they do take enough capercaillie eggs to cause a frown on a conservationist's brow. But above all, the species that I would most love to see here is a species that I can see every day of the summer without leaving my house in Northamptonshire. It is the tree-nesting swifts of Abernethy that fascinate me and which I have never seen. The swift is a familiar species to us all because it nests in our buildings and is a commonplace throughout the summer months. Parties of swifts scream up and down the road in which I live all summer – or at least from the very end of April through to the last lingering sightings, usually had on a warm summer evening with a glass of Rioja in hand at around the time of the Bird Fair. But really, swifts belong in places like Abernethy – where else did they nest, before buildings, other than in ancient woodland? The handful of swifts nesting in the ancient pines of Abernethy, seem to me to be a stamp of approval on the place, verifying its primordial nature and its high value. Yes, swifts are more commonplace than ospreys, but the uncommon nesting of those swifts in Abernethy tells us a lot about the ancient heritage of its amazing wildlife.

Lessons learned

Nature reserves are a nature conservation organisation's footprint on reality. They are the real world with all its problems and challenges, and they are the places where hard decisions have to be made.

Buying land is an expensive business and the costs don't stop at purchase because management costs are often high and are forever. And although land ownership doesn't give you absolute power over all the things that can affect the delivery of your nature conservation objectives, and despite the fact that often it takes a long time to turn a nature reserve from an infant into a fully rounded adult, there is nothing quite like land ownership to protect the nature of a place – if you get it right!

Land ownership is a 'safe' bet for nature conservation because you can always achieve something if you own enough land – it's just a pity that the price of land is so high. I can't think of many mistakes that the RSPB has made in land acquisition – there was only one nature reserve that Graham Wynne ever really regretted buying and he even changed his mind on that one over time!

Big nature reserves are better than small ones for a whole variety of reasons to do with ecology and economics but very few UK nature reserves are big enough – most would benefit from a doubling in area and if you have spare cash at the end of your life then please think of leaving some of it to a nature conservation organisation to buy land and manage it.

I find it difficult to get away from the thought that the world is a crowded place and that the more land that can be set aside for nature now, the better. But we can't buy everything we would like and even if we could there are some problems that can't be solved by owning land.

> *As soon as the land of any country has all become private property, the landlords, like all other men, love to reap where they never sowed, and demand a rent even for its natural produce.*
> **Adam Smith**

CHAPTER 10

Climate

We're finally going to get the bill for the Industrial Age. If the projections are right, it's going to be a big one: the ecological collapse of the planet.
Jeremy Rifkin

The RSPB started as a campaigning body – campaigning about individual acts of nastiness to individual birds when species such as egrets, terns and ospreys were slaughtered for their plumage. The Society has come a long way since then, to the point where RSPB staff attend international climate change talks across the world (with the expenditure of all that carbon) and argue the case for urgent reductions in greenhouse gas emissions because a changing climate is bad for wildlife (and for people). That's a big journey – is it the right one?

I played a large part in getting the RSPB into the climate change issue all of a dozen or more years ago. Back then it felt like an area which we couldn't avoid – despite the fact that at the time some staff argued that it was a step too far. We couldn't make a difference. It was just the next trendy thing and would pass eventually. I am sure that climate is an issue in which the RSPB should be involved. It has huge implications for birds and all our wildlife, so how can conservation organisations stand aside?

There is another reason why we can't really avoid the subject. The RSPB is proud of its one million plus members but a consequence of those large numbers is that about two percent of UK greenhouse gas emissions are due to the behaviour of the RSPB membership. A tricky area. Particularly because the RSPB's members will hold the whole range of views as to whether climate change is happening, is caused by human behaviour, is important, can be influenced, etc. In fact, I would guess the RSPB membership contains a slightly higher proportion of climate sceptics than does the population at large.

This is not the place to run through the evidence for climate change being real, and really being caused by our emissions of polluting gases, and nor am I the best person to rehearse that evidence anyway. So I will take climate change as read and spell out some of its consequences for wildlife before talking about what, if anything, we should do about it.

Nature on the move

My hugely clever colleague, Professor Rhys Green, worked with academics at Durham University (Brian Huntley, Yvonne Collingham and Stephen Willis) to predict, or at least suggest, what the future ranges of bird species might be like under a new climate. This analysis depends on the idea that current bird distributions are well-explained by climate factors – something which clever bird people used to deny, but now believe. But if the northern limit of the breeding distribution of, say, the robin is determined by how cold the winters are, how dry the summers are, etc., then by plugging in the available future climate scenarios one can see where robins might live in the future. That's the logic and it's quite straightforward.

Doing it though is a mammoth task! Luckily – though luck played little part in it – bird distributions are well documented so the raw data exist. Rhys and his collaborators took the whole of Europe as their study area and looked at all the bird species breeding there. Individual mathematical models were then created for each species and the climate extrapolations done. Simple!

And it was all written up and published in a book – *A Climatic Atlas of European Breeding Birds* (published in 2007). It is not, perhaps, the most riveting read of all time – lots of maps of potential range shifts in birds – but is a fascinating tome nonetheless, and I'm glad that I have a copy of it to dip into now and again.

Overall, not surprisingly, species in our hemisphere are 'predicted' to move further north as the climate warms. But the scale of this movement is surprisingly large - an average of 350 km northwards by 2080. And overall, the average European bird species will lose about 20% of its range by area in that time – partly because there is less and less Earth surface as you move towards the poles and partly because the proportion of sea to land increases.

But I put 'predicted' in quotations as that isn't exactly what this study does. It really says that habitat distributions will respond to changes in climate and that birds will follow their favoured habitats. So, if, and only if, patches of the right habitat exist and the birds can find them, then, and only then, would species be able to shift their distributions in response to changes in climate. It's not saying they will – it's saying they might.

I always think of the red grouse when this subject comes up. As a species at the southerly end of its world distribution in this country, the model unsurprisingly predicts that its range will shift northwards and that towards the end of the century it may be lost from England and Wales (time to sell your shares in grouse shooting?). It may be that the loss, or near loss, of grouse from Dartmoor and Exmoor in my lifetime is a sign that climate change effects are already happening. But the Climate Atlas prediction for red grouse is that the nearest suitable new habitat for them in 2080 would be in Iceland. Iceland – that's a long way away. Can you imagine red grouse making it from their nearest breeding grounds (Scotland and Northern Ireland) to arrive safe and sound and set up shop in Iceland? Maybe – but quite possibly not.

The Dartford warbler is another fascinating example. It will definitely spread north in the UK – it already is – but we all know that it is highly dependent on heathland habitat and so will only spread into heathland and moorland areas. Towards the end of the century the Dartford warbler might well be common in the Peak District and the North York Moors. But it will be rarer in Spain and Portugal – where it is now common. Climate changes will lead to shifts northwards (and up-the-hill-wards) but that will apply to the southern edges of species' ranges too. Central Spain will have a different climate – more like north Africa – and that won't suit the Dartford warbler, we think.

Take another example, the massive decline of the willow tit in the UK is very striking. It used to be common in England south of the Midlands, which is why Bert Axell knew it at Minsmere and Jeremy Sorensen sent me out looking for it, but is now very difficult to see there. All sorts of reasons might be behind this decline, but the fact that its current distribution is approaching that predicted for a more distant time with a changed climate, makes me wonder whether those climate impacts are already beginning to bite in ways that we don't fully understand.

It's not just breeding species that appear to be affected. Many wintering species are showing signs of change. Some, such as European white-fronted geese, whose flocks I used to scour for rarer species at Slimbridge as a lad, are much less common in the UK than 40 years ago. But white-fronted geese have increased in numbers in Europe – it's just that fewer and fewer of them are bothering to come as far west as the UK because, we think, milder winters in continental Europe don't force them to move. The other star-species at Slimbridge, the Bewick's swan, has also declined in the UK but that situation is also reflected over the rest of its European population. These examples show the real importance of looking at the big picture rather than just that little bit of it around the Severn Estuary, southwest England or even the UK as a whole – birds are mobile and operate on a large scale.

A fractured ecology?

So species are on the move already, and they will move further as climate change progresses. But these spatial movements are accompanied by temporal ones too. The timing of events is changing. Insect emergence dates are changing, birds' nesting seasons are changing, arrival dates of migrants are changing – the natural world is in a state of flux. The big worry is that different species react in different ways and speeds, and that intricate food webs are becoming stretched to breaking point.

Again, there is already evidence that this is happening. Pied flycatchers in the Netherlands depend for successful breeding on their young being raised when the abundance of their caterpillar food peaks. Caterpillars are now most abundant earlier in the year than before, probably due to climate change, and pied flycatchers have not advanced their breeding season to match – probably because they are constrained in how quickly they can return from their wintering grounds on the other side of the Sahara.

One of the big trends in bird populations across Europe now is that migrant species are declining at a greater rate than residents. This might be nothing to do with climate change of course – it might simply be that the African wintering grounds of many migrant species are being wrecked faster than we are degrading our own countryside, or that since migrants are likely to be more dependent on insects than are resident species it's something to do with insect numbers here in the UK and elsewhere in Europe, rather than what actually happens to those species in winter.

But there is a fascinating study by Anders Pape Møller, Diego Rubolini, and Esa Lehikoinen which certainly indicates that climate change factors may be involved. They found that when one looks at the European arrival dates of 100 species of migratory birds, such as warblers, hirundines, flycatchers, etc., then most species now arrive earlier in the year – by a few days or even some weeks. The really interesting thing is that those species which have shifted their arrival times the most are the ones that have declined the least. In contrast species which have not changed their migration in this way, and arrive at the 'usual' time, are those which have declined the most (e.g. wood warbler). One explanation of this pattern would be that climate change has influenced the timing of emergence of insects and that only those species which can return to their breeding grounds earlier can adapt to these new conditions. It's not the only explanation and it's certainly not proven, but it is an attractive idea which would explain lots of things.

It looks to me as though we are carrying out a big experiment on the planet – we are heating it up without knowing the consequences. Maybe nature can cope – maybe it can't. At present scientists estimate that a third of the species now living on Earth may be committed to extinction by the impacts of climate change by the year 2050 unless we change our lifestyle. Maybe we can cope – maybe we won't cope so well. But, as is often the case, the way we live our lives imposes consequences on millions of other species with which we share this planet – although we 'share' it in a rather selfish way. And, as is also often the case, the impacts of our actions on nature tell us something about how we ourselves may be affected. Such predicted major changes in the distributions and survival chances of the species around us ought to give us pause for thought. A bit of a decline in pied flycatchers may not worry you too much. However, a bit of a decline in pied flycatchers alongside declines in scores of other species, early evidence of fracturing food chains and predictions of major shifts in species' distributions is quite a big deal. It links a trivial (-ish) observation in your local wood to major changes on the planet. If climate change is the common cause then it is a big deal – a really big deal. It's a big experiment and nature plays the canary in the coal mine – except with this one, there's no way out of the coal mine and we're all in it together.

So what, if anything, can we do about it? The answers are of two types, which in the jargon of the subject are called adaptation and mitigation. Adaptation is what we might do to adapt our ways of life to the new climate which is inevitably coming (inevitably because greenhouse gases stay up in the atmosphere for decades so the stuff we sent up there decades ago is still exerting an effect). Mitigation is changing our habits now in terms of the causes of climate change so that the future impacts are reduced.

Adapting to a new climate

By 2050 our grandchildren will live in a UK (if that is still what we call this little country of ours) which is warmer overall by at least 2°C, drier in summer and wetter in winter (broadly speaking). Extreme weather events (floods and storms) will be more common and the sea level may have risen by about 0.5 m. This will have consequences ranging from you having to cut your lawn all through the winter to needing air-conditioning in our cars, homes, hospitals, schools and workplaces. Farmers will grow different crops or may need to store water for irrigation over the summer – but maybe that stored water will be needed for drinking too. London will be increasingly threatened by rising seas and expensive barriers will have to be built. The uplands of Britain may become less harsh environments and more suitable for different forms of agriculture and might support a greater human population.

How will nature be getting on? One of the conservationists' worries is how to plan management of nature reserves in this uncertain future. Certainly differently, but how exactly? Coastal nature reserves such as Titchwell and Minsmere will be threatened by rising seas – they already are. At Titchwell major work on realignment has recently been completed and at Minsmere the fresh marshes which provide a home to bitterns are threatened every time that high tides and high winds combine to create a surge tide. How does one adapt to sea level rise when one is on the coast?

There are two main options – build better defences or move away! The RSPB is doing both. The creation of Lakenheath Fen as a reedbed was a conscious decision to create bittern-ready habitat that could in some way 'replace' coastal sites such as Minsmere, Walberswick, Easton Broad, Benacre Broad, Titchwell and others in time.

Many RSPB nature reserves are wetland sites. Will there be sufficient water available in future decades? Will the market price it at too high a level? Will there be enough water for wetlands such as the Norfolk Broads and the Somerset Levels to survive? Difficult issues which will be determined not only by whether scientists' predictions of the future are correct but also by how the rest of human life changes to adapt to a new climate. Will our own selfish responses give nature a fair deal in the future?

We already see 'new' species arriving in southern England. In 2010 purple herons bred for the first time in recorded history (at the RSPB Dungeness nature reserve) and little bitterns for only the second time in living memory at the RSPB Ham Wall nature reserve. Red-backed shrikes nested on Dartmoor in the same year, and again in 2011. We are now very accustomed to seeing little egrets in places where only a couple of decades ago we would never have seen them. These are all southern species which have spread to the UK and they are accompanied by many insects behaving in the same way.

One of the RSPB's newest nature reserves is in east Kent near Sandwich Bay and will, when fully restored, be an amazing wetland perfectly placed to accept new arrivals to the UK – will they be fan-tailed warblers, penduline tits, bluethroats, great reed warblers or some other unthought-of species? We'll see! But the RSPB thought ahead and decided to re-create a wetland there because of its present value but also its strategic position in a world of changing climate.

At the other end of the country, back in my beloved Flow Country, the greenshanks, divers and common scoters may find the climate changing in ways to their disadvantage. As will the whimbrels and phalaropes on Shetland. Dotterels, moving further up their mountains in search of an alpine climate, will run out of suitable space too perhaps. How will these species adapt and what should we do to help them?

And in between, when are we going to start creating the habitat links that will enable Dartford warblers to spread north to the Scottish Border – and how will the heathland smooth snakes and sand lizards make the same journey? These may all seem like questions for the future but the RSPB and other conservationists are thinking about them now – and they need to be thought about now, if the solution is planting a wood or recreating a heath which will take decades to come to fruition.

If only it weren't going to happen! Well, unfortunately climate change *is* happening and it *is* our fault – 'our' in the sense of our species' fault. The changes in climate that future generations of people and wildlife will experience is determined by our – that means your and my – actions today.

Mitigation – can we make it less bad?

Mitigation is about making it less bad. How do we live our lives and yet pump fewer and fewer greenhouse gases up into the atmosphere to affect all life on Earth for decades to come?

Let's just be clear about which gases we are talking about and where they come from. Water vapour is actually the most potent greenhouse gas but our activities haven't altered its amount in the atmosphere significantly. It's important in determining the temperature of the planet but not in changing that temperature. The main greenhouse gas produced by humans is carbon dioxide, mostly resulting from the burning of fossil fuels such as coal and gas. Then there are methane (largely from the burps of ruminants) and nitrous oxide (a major by-product of inorganic fertiliser).

A carbon dioxide molecule (CO_2 – one atom of carbon joined to two oxygen atoms) remains in the atmosphere for tens of thousands of years whereas a methane molecule (CH_4 – a carbon atom joined to four hydrogen atoms) only stays up there for about a dozen years and a nitrous oxide molecule (N_2O – two nitrogen atoms joined to one oxygen atom) stays above our heads for about a century. By measuring the concentrations of these gases in air bubbles in ice cores and elsewhere, we know that the concentrations of these three greenhouse gases, CO_2, CH_4 and N_2O have increased in the atmosphere by about 30%, 66% and 16%, respectively, since the start of the industrial revolution. That's why the planet is cooking.

The potency of the greenhouse effect is such that if we removed the water vapour and other greenhouse gases from the atmosphere, the Earth would be 33°C cooler – that would make the planet a very different place. But the impact of the increased greenhouse gas concentrations is such that the planet is likely to warm by 2°C almost whatever we do – and by 4–6°C by 2050 if we don't do something. Two, four and six are small numbers but that doesn't mean that their impacts will be small – inland floods, droughts, inundation of coastal areas, mass migrations of people, huge changes in crop yield will all affect how much food could be grown and whereabouts people could live. The poor will, as always, be affected more than the rich but the world will be a very different place – and a much nastier one.

Who put all those greenhouse gases into the atmosphere to start with? Over the last century the USA has been the chief climate polluter, accounting on its own for almost a third of greenhouse gas emissions, with another four countries or regions accounting for the next third (the EU 23%, China 8%, Japan 4% and India 2%) and the rest of the world accounting for the remainder.

At the present time, the countries topping the list for today's emissions (2005 values) are led by China (17%), the USA (16%), the EU (11%), Indonesia (6%) and India (5%).

Scientists suggest that we need to reduce our greenhouse gas emissions by 80% over the next 40 years if we are to stand a good chance of limiting the rise in global temperature to a mere 2°C. Eighty percent? – that's huge! How would you manage on only one fifth the travel, or electricity or fuel use of today? To say that it would be tricky is such an understatement as to be ridiculous.

So, almost all the routes to reducing our greenhouse gas emissions are either inconvenient or apparently impossible. Here are a few things we could do: cover the country in wind turbines, build barrages across estuaries, stop flying, stop driving, stop eating meat, build nuclear power stations, insulate our houses better, fit solar panels, and perfect the design of electric cars.

And the truth is – you won't get very close to that 80% reduction by choosing any one of those options – you'll need a combination of them to make much impact.

The last Labour Government brought in a UK Climate Change Act with a fair degree of cross-party support that does, indeed, commit us to reducing our greenhouse gas emissions by 80% by 2050. The Climate Change Committee is working on the details.

Now that I'm not an RSPB employee I can tell you what *my* preferred options would be – I'd do some of all of those options above. And that includes building new nuclear power stations, an issue that I tried very hard to keep the RSPB out of whilst I was Conservation Director. I could see nothing to be gained by the RSPB entering the arguments about nuclear power – the conversation is a crowded one already and we would have added little to the debate. But whilst I do know the dangers of nuclear power I think we have to set them against the dangers of climate change. The trouble with climate change is that it affects everyone.

Now whether or not you are going to build new nuclear power stations you will need more renewable energy generation too. Wind energy is a good option in the UK but it certainly isn't popular with all. A single wind turbine won't produce that much green energy just as a single field of wheat won't feed many people. If wind energy is to make a difference we will need lots of turbines and they will become a much more familiar part of the landscape. So, just as I can see ten wind turbines from my house in east Northamptonshire, more people may need to get used to these intrusions of industrial architecture into their rural landscapes (and urban ones too) if we are to head towards 80% greenhouse gas reduction.

The RSPB has had a love-hate relationship with windfarms over the years – not switching from love to hate and back again, but holding those two emotions all the time. We love wind turbines because, despite what you might read elsewhere, they can make an important difference to our greenhouse gas emissions, but we hate them if they are put in the 'wrong' places. Our version of 'wrong' is in places where they will do harm to birds or other wildlife – such as in protected areas for birds on land or sea.

The evidence that wind turbines can cause harm to birds comes mostly from abroad. There is good evidence from the USA and Spain, particularly from migration hotspots, that large windfarms can kill lots of eagles and vultures. A windfarm erected on the island of Smolo off the Norwegian coast is thought to have wiped out a large population of more than 30 pairs of white-tailed eagles over a period of just a few years. There is much that we don't know about the impacts of wind turbines on birds but white-tailed eagles do seem to be a vulnerable species.

The RSPB has worked with windfarm companies to change the location and design of windfarms so that their impacts will be lessened or removed and sometimes we have campaigned against them, such as with the proposal to build a windfarm on the island of Lewis.

But if you don't want nuclear power and you don't want windfarms then maybe you'll give up flying and your car? Few of us want to do anything like this. You can see why the truth of climate change is indeed inconvenient.

Looking at greenhouse gas emissions on a per capita basis, rather than by country, the average UK citizen is responsible for about 11 tonnes of CO_2 equivalent emissions each year (taking N_2O and CH_4 into account as well). That puts us in the top one-sixth of countries. Your average American friend is more than twice as polluting as you are, the average German a bit more polluting and the average Frenchman a bit less polluting (those nuclear power stations are helping!). The average Chinaman is now only about half as polluting as you are and he is catching you up fast, but the average Indian has only now become one sixth as polluting as you. At least the UK has reduced its emissions a bit in real terms in recent years.

There are various places where you can assess your own carbon footprint to see where you stand in the UK list – are you above or below the UK average? If you are around average then your direct carbon footprint will be dominated by travel – car use, public transport and flights. 'Going places' will account for about a third of your total emissions. Then will come heating your home and water, and using appliances and lighting in the home. These 'switching things on' activities will account for about another third. And the final third is embedded in the goods you buy and the food you eat – somebody else drove your food around before you bought it and built the road that you travel on, and your share of that carbon is assigned to you! Sorry – that's how it works.

Looking at your own carbon footprint does make this seem very personal – that's a good thing, I think. But you aren't in charge of how your electricity is produced so you have to live with the carbon footprint of your energy supplier. Except, of course, if you decide to exercise some control yourself. First, you can choose to get your energy from a green supplier – although you may well be encouraging the need for a windfarm to be built outside your window or where it can be seen from a National Park. Alternatively (or in concert), you can vote for politicians who will actually do something about decarbonising (horrible word – but can't think of a better one) energy supplies (perhaps by building some more potentially disastrous nuclear power stations).

While you aren't completely in charge of your travel emissions, you can exert a good deal of influence. Depending on your circumstances you may or may not need a car – and it does depend on your definition of 'need'. Some cars are vastly more fuel efficient than others, so your choice of car is important here – you can reduce your emissions with a bit of careful choice. The Ford Zetec I drive is not the most fuel efficient car in the world but it's not bad at all – and given that I 'needed' a car with enough headroom for a six foot three inch me, and enough room for a family of two grown up kids and the occasional granny as well, then it does what I need and produces fewer of the greenhouse gas emissions that the planet does not need than many other cars I could have bought.

And if I cut down how much I drive (which I have) and drive more economically (which I now do) then that makes further significant savings.

How about those birding trips abroad? What about planes? I gave up travel on planes for recreation six years before I left the RSPB – and then cracked and treated myself to a trip to the USA! At work I turned down many opportunities to travel which I could have taken – I didn't go to the BirdLife International World Congress in Argentina even though I would have loved to have gone (and seen some new birds). I turned down all-expenses-paid trips to Korea and Greece to talk at conferences and I had protracted discussions with conference organisers about whether they would pay my train fare to Glasgow rather than the cheaper flight costs! But sometimes you just have to fly.

In September 2009 I flew to Northern Ireland to do some interviews. I wrote in my blog as follows:

> *My share of the round trip flight was around 150 kg of emitted carbon dioxide. So that's over my own weight in CO_2 – even though I could do with losing a few pounds (or stones actually!).'* and '... *my return flight was about one and a half percent of the average annual emissions of a UK citizen. By comparison - just that short return flight is equivalent to about the whole per capita annual CO_2 emissions of citizens of about 20 nations on Earth - mostly African countries. The average Tanzanian or Ugandan emits less CO_2 in a year than I did in just over a day - and we aren't counting my drive to the airport and then home, the TV and lights in the room, the energy that went into heating the water for my shower or cooking the food for my meals.*
>
> *My father never got on a plane in his life, my Mum was in her 70s when she first flew. I was in my early 20s when I first got on a plane and my children were under 10 on their first flights. That might be progress - it certainly felt like it at the time. But if greenhouse gases are wrecking the climate then it is difficult to see flying as an absolute and unquestionable right.*

That is how I still feel. Each time you get on a plane you are wrecking the climate just a little bit more. If you economise elsewhere in your carbon budget – drive less, turn the heating down and lobby government for a decarbonised economy then you can occasionally get on a plane and not feel guilty, but if you get on a plane and don't give it a thought then you are acting as an irresponsible inhabitant of the planet. The truth is inconvenient.

How about your food? There are plenty of carbon tonnes tied up in that too. But the main greenhouse gases associated with farming are nitrous oxide and methane. Ruminant animals, such as sheep and cows (but not pigs or poultry), produce large amounts of methane through their digestion. Some of this comes out of their backsides, but most comes out of their mouths – each time a cow burps it helps to wreck the planet. Does that make you feel less guilty about your plane use I wonder? Cows are polluting the atmosphere all the time, even when sitting down! But as with most other things, don't think there is an easy solution – like switching to eating pork and poultry. Although these sources of meat do not, while still alive, have the same high methane emissions as ruminants, they are associated with other emissions that can make you feel guilty. Soya is a major component of the animal feed imported into the UK, much of which goes to feed pigs and poultry. Much of that soya is grown on land that used to be rainforest and the loss of that rainforest, and the carbon that went into the atmosphere as a result, has to be on someone's conscience – or at least on their national or per capita carbon account. Where do we put it? Does it accrue to the country where the rainforest was cut down – perhaps Brazil or Indonesia – or does it belong with you as the end consumer? Obviously, the atmosphere has taken the same carbon hit wherever we, accountant-like, add it to the books. The answer probably has to be a bit of both.

What about those nitrous oxide emissions associated with our food? They come largely from the use of inorganic fertilisers to grow crops such as wheat or to improve grass growth. Some of the fertiliser is taken up by the crop, some ends up in watercourses and some ends up in the atmosphere. Did you know that about 40% of the wheat grown in the UK goes to animal feed? So it's the nitrous oxide emitting wheat fields which are feeding the methane-burping cattle. I'll be telling you to be vegetarian next! Well…not exactly, but if you were vegan then your carbon footprint from food would be considerably lower. A vegan diet would have as much impact on your carbon footprint as switching from a gas-guzzling car to a hybrid. So it's well worth thinking about on climate grounds alone – but you will also be helping to stop rainforest destruction that will drive species to extinction, helping to reduce water pollution and freeing up more land for more efficient food production in a world where many still starve. So there's a lot to be said for the vegan lifestyle.

So am I vegan – no way! But I am vegetarian for four days a week. That's my equivalent of drastically reducing my air travel and cutting down on car use. I have cut down on food emissions too. And it really wasn't difficult to do. The most difficult thing is actually the social pressures that push you to eat meat – poor vegetarian choices in restaurants, not wanting to inconvenience friends, the lack of good labelling and the fact that whenever a choice of vegetarian and meaty food is provided all the meat-eaters tuck into the vegetarian food with gusto!

But the message, if you have stuck with me so far in this chapter, and I expect there will have been some reader-casualties along the way, is that you are not helpless when it comes to your carbon footprint. Some of what we do to the planet is beyond our control in the short term – although if you are politically active then you can make a difference in the long run. But there are plenty of areas where you can make personal choices which will reduce your carbon footprint. I've made some but I am not claiming saintly behaviour. What I would say is that each of the choices I have made in reducing my air travel, car use, energy use in the home and carbon footprint from food seemed more difficult in prospect than it did in retrospect. If you believe that climate change is a problem and you don't do something, then it is partly your fault! But if you do something then the feeling of reduced guilt and smugness is worth having. You can hold your head up with your friends and swap stories of carbon reduction. Just don't try it with the 80% of the world's population whose carbon footprint is lower than yours or mine.

Climate change is the big environmental issue of the age. As a naturalist you will see evidence of it in the world around you in a way that others may not: the earlier arrival of some migrant birds and the creeping forward of birds' breeding seasons; the more northerly distribution of southern species such as the little egret; the disappearance of some more northern species such as willow tit from your local woods; the increasing vulnerability of your favourite coastal nature reserve to sea level rise and storms. The more in tune you are with nature the more you will notice the changes and want to help nature to cope with, or adapt to them. And I think the more in tune you are with nature, the more you will want to do in your own life to lessen the scale of those changes.

Conclusion

Climate change will ruin this planet for all life – the wildlife we love and the people it sustains, unless we do something about it. 'We' in this case means the governments we elect, through their policy decisions, and ourselves as individuals, through our personal lifestyle choices. This is not something that is being done to us by others – it's something that *we* are doing to future generations of life on this planet.

Nature is already on the move and we can see actual signs of the effects of climate change on the birdlife around us. Nature conservationists can either throw up their hands in despair or take decisions now on where to buy land, how to manage it and what else they can do to give wildlife a helping hand to redistribute itself over the surface of the Earth, and survive.

At heart though, climate change is just one symptom of our unsustainable way of life. It's a pity that greenhouse gases have such a long life in the atmosphere because we are living with the 'mistakes' of our forebears who were less well-informed than we are now and unless we act now, our children will inherit these mistakes – plus our own.

Climate change is *the* big test for our species. We know the causes, we know the consequences, we know some of the solutions – but will we take action?

> *In the year 2065, on current trends, damage from climate change will exceed global GDP*
> **Andrew Dlugolecki**

CHAPTER 11

The raptor haters

Whether or not Hen Harriers continue to inhabit this world is a matter of indifference to the great majority of mankind.
Donald Watson

For a nation of animal lovers and a law-abiding people, there is an awful lot of wildlife crime going on out there and a lot of it is directed against birds of prey.

I have always believed that illegal killing of birds of prey is just one of those core areas where the RSPB has to take a principled stand and say, loudly and often, that this illegal activity is wrong – deeply wrong. If the RSPB isn't saying that then who will?

Most of the killing of birds of prey emanates from the shooting community – not that everyone who shoots is killing birds of prey by any means, but most birds of prey which are illegally killed are almost certainly killed deliberately or carelessly by shooting estates. So rather perversely it may seem, I will start this chapter by explaining how birds of prey, and most particularly one of my favourite birds, the hen harrier, can be seen as a pest by those who manage driven grouse moors and how those feelings spill over into generalised raptor-hating in parts of the shooting community.

Hen Harrier

In Chapter 5 I described how driven grouse shooting, legally practised, involves an awful lot of work and investment before the 12 August when the grouse season opens. There is the heather-burning, legal predator control and provision of all that medicated grit to be seen to. All this effort culminates in having large numbers of red grouse that can be chased across a moor so that they fly at full speed over a line of guns. You can see that a shooting estate charging hundreds or thousands of pounds for a day's shooting in order to recoup their investment and turn a profit in the good years is not going to be very keen if a lot of those red grouse disappear into the stomachs of the local golden eagles or hen harriers.

The Langholm study

Long ago, we in the RSPB used to hold, in all honesty, to the view that a few hen harriers scattered around the uplands weren't going to present a great threat to driven grouse shooting and that the grouse moor managers' fears were greatly exaggerated. We were wrong.

It was the Langholm Study, more properly called the Joint Raptor Study, that showed how wrong we were.

I remember the Game Conservancy's Dick Potts talking to me about whether the RSPB would join in and help to fund the study at Langholm in the Scottish Borders. Dick told me that the study he and others had in mind would involve 'protecting' the hen harriers on a Scottish estate and seeing what happened when harrier numbers increased – as presumably they would once fully protected. How many harriers would there be and what impact would they have on the shootable surplus of red grouse in the summer?

As I remember it, Dick and I thought that harrier numbers might increase to maybe half a dozen pairs on the Langholm Moor estate of the Duke of Buccleugh and that that might mean that the shootable surplus of red grouse might halve in numbers or thereabouts. What actually happened took just about everyone by surprise. Hen harrier numbers rose to over 20 nesting females and they were eating so many grouse that there were not enough even to have a grouse shoot that autumn. During the summer the harriers had simply eaten all the birds that would normally be available for shooting at the end of the summer – terribly unsporting of them!

So I can see that if you are a grouse moor manager you would regard hen harriers simply as a pest. You would wonder why, if you can shoot crows and magpies you can't shoot hen harriers too. On many, but not quite all, moral grounds you could be right about the equivalence, although how that should shape your view on predators in general might be open to further discussion.

The fact that something is illegal, and hen harriers have been fully protected by law since 1954, does not necessarily mean that people will stop doing it. I have a feeling that I have exceeded the speed limit once or twice in my life and I expect you have too. Raptor killing is not irrational even though it is illegal – if you are operating in a business which depends on killing lots of crows, foxes and stoats to stay in the business of killing lots of red grouse then I can see why killing a very few hen harriers doesn't seem like too big a deal, and if you aren't going to get caught when you do then I can see why many are tempted. These days it seems that much of the killing is done away from the moors at winter roosts.

Perhaps this wouldn't matter so much, and perhaps hen harriers wouldn't even have the level of legal protection that they do, if they were more like crows and foxes in population terms. Whereas no gamekeeper will do himself out of a job by killing so many foxes that he never has to kill one again, the same cannot be said of hen harriers. Hen harriers were pretty much wiped out in the UK mainland before the Second World War and it was only the relaxation of persecution during the war years, legal remember at the time, which allowed them to make a recovery.

Hen harriers are a pretty rare bird with around 800 pairs across the UK, so they are a very different kettle of fish from the crow or fox. While 800 pairs doesn't sound too bad at first, remember the scientific estimates of the likely population in the absence of illegal persecution – c. 2,500 pairs – and you realise what we are all missing.

The hen harrier is now doing reasonably well in those areas without grouse moors and very, very badly in all those areas where grouse moors are the predominant land use. In 2011, there were only four successful hen harrier nests in the whole of the north of England whereas one would expect more like 250–300 pairs, based on the habitat available. That's an awful lot of missing birds. Hen harriers are also missing from those parts of south and east Scotland where grouse moors are a common land use. In contrast, in the Isle of Man, Wales, northwest Scotland and Northern Ireland – all places where driven grouse shooting is rare or absent – hen harriers are doing reasonably well.

It appears to be an intractable problem and certainly not much progress has been made in recent years. Hen harriers are protected by law but cause real problems for grouse moor managers. Is there a solution to be found?

Harrier solutions

To my mind there are three potential ways forward for the nature conservation side of this debate – which essentially means the RSPB and a bunch of raptor enthusiasts.

Route 1 is simply to give up, go off and do something else instead. Hen harriers aren't a highly threatened species in global terms and their numbers are not very low. Yes they are absent from large areas of suitable habitat, and yes that's because of illegal activity by shooting interests but maybe there are other things more important, more tractable and more worthwhile to do.

This may be the route already chosen by some nature conservation organisations. The National Trust owns a lot of upland land which should have hen harriers flying around it, delighting their visitors, but the National Trust does not shout very loudly on this subject. The Wildlife Trusts don't say much about it either. That leaves the RSPB to decide whether it should go down Route 1 as well. Personally, I don't think that would be an honourable course. I never really considered it in the past and would think very badly of the RSPB if it took it now – and I'm fairly sure it won't.

Route 3 (we'll obviously return to Route 2 in a moment) would be quite an exciting journey and that would be to campaign against driven grouse shooting on the grounds that all reasonable avenues have been explored and led nowhere so the 'nuclear option' is the only choice. The RSPB's Royal Charter makes things a bit tricky in this respect, but if you really believe that widespread and illegal persecution of hen harriers by those engaged in grouse shooting is the reason why the population of this bird is a third of what it should be, then clearly this illegal practice has a significant conservation impact and it is easy to argue that grouse shooting is far from a legitimate field sport – it is a field sport dependent on illegal activity.

Under these circumstances it would be perfectly possible to forge a coalition between the RSPB, the League Against Cruel Sports, the RSPCA and raptor enthusiasts to mount a campaign to ban driven grouse shooting. Such a campaign would probably gather great public support. Once, though, you get into the nuclear option then it is impossible to make progress with the good guys and the die is set for decades to come. Parallels with the fox hunting ban are legion.

Route 2, the middle way, will appeal to the British love of compromise – and that may be no bad thing. Research at Langholm Moor suggests that supplementary feeding of hen harriers during the breeding season can greatly reduce the number of grouse that they eat – why bother hunting if someone is providing you with freshly killed rats, day old chicks or venison to eat instead? This result has been tested at Glen Tanar in Grampian, as well as at Langholm itself, and the results look promising – very promising. Food provisioning prevents losses of red grouse and is probably a technique that could be employed much more widely. And if hen harriers are cosseted on grouse moors in future – and that would certainly be a dramatic change from the current situation – it would mean that their numbers would rise towards that 2,500 figure. More and more grouse moors wouldn't just have the odd pair of hen harriers – more and more would have many pairs.

Given the situation, I can quite understand that grouse moor owners as a group would quite like an escape route. They have suggested that they be allowed to cap hen harrier populations on any moor at a set level, such that driven grouse shooting could continue. The capping could be done by shooting birds, removing eggs or perhaps by taking young harriers from nests to release them on other, distant, moorland areas.

Which Route would you favour, if any? Personally I dislike Route 1 but would consider either Route 2 or Route 3. Maybe Route 3 should be held in reserve, but rolled out if Route 2 fails – and there is no certainty that it would win favour with many grouse moor managers. After all, enough grouse moors appear to be happy enough on their own route, Route 4, which is 'We're quite happy to carry on breaking the law and you can't catch us'.

It should be said that the management of grouse moors should be closely examined for lots of other reasons too. Is it really as good for biodiversity as grouse moor managers claim? Is the economic value of grouse shooting really that important in upland communities? Remember, it's not just the hen harrier that is missing from grouse moors – golden eagles and peregrines are also absent from lots of sites.

It isn't as though these matters haven't been talked about many times in private and in public. It isn't that there has been no attempt to find common ground – it's just that no common ground has been found. Years ago I stuck my neck out in a private meeting and suggested to grouse moor managers that if hen harrier numbers in England reached 40 pairs then, although that was far below the 250–300 pairs that there 'should' be, the RSPB would ease off on the subject of hen harrier persecution. At the time I thought I was being quite brave, and quite possibly betraying the hen harrier, but despite immediate recognition that my offer was a step forward there has been no improvement in the status of the hen harrier in England – quite the opposite.

I also wrote a piece, published in the *Shooting Times*, which asked those good men in the grouse shooting community to stand up and work with nature conservationists against those, perhaps few, who break the law. Although that initiative was well-received at the time no progress was made then either. So it certainly isn't unreasonable for the RSPB and others to consider the nuclear option now. There has been plenty of jaw-jaw to no avail.

I do think that it's time to break the stalemate as it certainly doesn't suit the hen harrier, golden eagle or peregrine. As time goes on I am more and more drawn to the nuclear option as the grouse moor community refuses to act in good faith.

Sandringham

So to the events on the Sandringham Estate of autumn 2007. But first let's remember that the estate has 'previous'. On 3 November 2006 one of the 14 gamekeepers on the Sandringham Estate was fined £500 (and £470 costs) after pleading guilty to setting an illegal trap which maimed a tawny owl. The gamekeeper had left a steel spring-trap, which he said was set for rats, uncovered which put birds at risk. A local farmer discovered the owl with its leg caught in the trap which was tied to a post. The RSPCA were called but despite veterinary treatment the owl had to be euthanased.

The defence lawyer said that it was the Sandringham Estate's policy to deal with vermin in a manner that was considerate and in compliance with the regulations. The presiding judge described the 'keeper's actions as 'unwise and unprofessional'. Ian West, Head of Investigations for the RSPB, said: 'High standards are expected of people working as professional gamekeepers on estates. It is particularly disappointing that this happened on a royal estate.'

But none of this prepared the world for the events of late October and early November 2007.

I was on a train to London on 24 October 2007 when one of my staff telephoned and said he'd had a call from a Natural England staff member who had just seen two hen harriers shot out of the air on or near the Sandringham Estate. I think the caller wanted to make sure that I wasn't going to say 'Don't get involved' and to alert me to the fact that this might be a high-profile case! But, of course, the RSPB's role should be to help the police in their enquiries into wildlife crime, when asked. We have valuable expertise that the police don't have. These days we might call it 'Big Society', I guess.

Mark Thomas' account of the enquiry on the RSPB *Investigations* blog made fascinating reading.

Here are a few extracts:

> *The evening of Wednesday 24 October is one that will live with me for a very long time. I had just reached home when my mobile*

phone rang with a call from the Natural England head warden at Dersingham Bog National Nature Reserve, a contact of mine through Operation Compass the anti-egg collecting initiative.

He swiftly gave me a mobile number of one of his employees and asked me to call him immediately as he had just witnessed the gunning down of two hen harriers on the Sandringham Estate.

Hardly able to speak, I rang the number, introduced myself and asked what was happening… He explained in a calm, yet clearly traumatised voice that barely 20 minutes ago, together with two members of the public who he had taken specifically to look for roosting raptors, he had watched two female hen harriers being shot out of the sky.

He explained that it was with great excitement that he had pointed the birds out to the young child present and that they were all watching the birds, two females flying in tandem over the edge of the reserve and onto the Sandringham Estate.

Suddenly, all heard a loud shot and the first bird folded up and plummeted to the ground. This shot was then followed immediately by a second and the second bird fell like a stone. A third shot was then heard shortly afterwards, and being a shooting man, the warden said he believed this was a 'finishing off' shot.

Just listening to this I felt physically sick, but it was not the first time I had heard first hand accounts of raptor killing and I focused on the job in hand. Even down the mobile phone, I could hear the blast of shotguns in the background and I was informed that the shooting was continuing, presumably legal wildfowling.

I took down all the details and immediately contacted a Police Wildlife Crime Officer at Norfolk Constabulary. Just like me, the Police Officer was extremely concerned and promised swift action. My telephone was red-hot all night and I finally went to bed around 1 am.

Early the next morning saw my colleague Guy Shorrock and myself heading through the quiet villages of Norfolk on our way to a Police briefing. Just after dawn, we were on the ground and being led by the Natural England warden to the scene of the previous evening's incident.

We entered the Sandringham Estate and after a few minutes arrived at a duck-shooting pond. A couple of people were already present: a man and a woman with a Landrover and eight dogs which were busily working the ground. On speaking to the people, we learned they were there on request of the estate to retrieve ducks shot the previous evening. They kindly agreed to allow the Police to search their vehicle.

Over the next few hours, a painstaking search of the area took place, but sadly no harrier bodies were found. We did locate a number of fresh, recently used lead shotgun cartridges at three points around the lake.

Being experienced investigators in these type of cases, Guy Shorrock and I had no doubt whatsoever that the warden had witnessed the illegal killing of the two hen harriers but that like so many cases before, these crimes are almost impossible to bring to court, particularly when the bodies of the victims are missing.

Later that day we had a debrief with a Detective Chief Inspector in the Norfolk Police. I was reassured by just how seriously he took the killing of hen harriers and the efforts he subsequently took to investigate matters, despite the unique circumstances in this case.

Ten days on, I am not surprised by the Crown Prosecution Service decision that there was insufficient evidence to bring any charges. The sick feeling in my stomach is still present, it just gets slightly less each day, until the next time my mobile rings…

The news actually broke on 26 October in the *Daily Mail* – an amazing newspaper whose views are often a bit mad but whose journalists seem to track down an awful lot of stories. The headline 'Who shot dead the Queen's rare harriers?' summed it all up. The story quoted Natural England as follows:

> *One of our wardens and the members of public saw two harriers shot and killed on Wednesday evening at about 6pm. They saw the birds in the air, heard the shots and watched them fall. It wasn't possible to see who was responsible. We are shocked that two of England's rarest birds have been killed in this way. They are one of the most vulnerable species we have.*

However, it was left to the *Guardian* several days later, on 31 October, to break what was the big story – that Prince Harry, the son of Prince Charles – had been interviewed by the police in connection with this event.

The *Guardian* reported:

> *Prince Harry and a close friend have been interviewed by police after two rare and legally protected birds of prey were killed on the royal family's Sandringham [E]state in Norfolk last week. The prince is understood to have been out shooting on the estate last Wednesday evening, with a friend believed to be from the Van Cutsem family, when witnesses saw two hen harriers in flight being shot, an offence under wildlife protection legislation which carries a prison sentence of up to six months or a £5,000 fine.*
>
> *Sources have told the Guardian that the prince and his friend were the only people known to be out shooting on the estate last Wednesday evening, and were quickly identified to Norfolk police by the Prince of Wales's staff. It is understood both men were interviewed in person, but have denied any involvement in the incident.*
>
> *Last night a spokeswoman for Clarence House said: 'Because Prince Harry and a friend were both in the area at the time, the police have been in contact with them, and asked them if they have any information that could help. Unfortunately, they've no knowledge of the alleged incident.'*

The media had a field day over this story. Catching up on the *Guardian*, the *Independent* wrote on 1 November: 'Again, how Prince Harry's grandmother can be both patron of the National Society [*sic*] for the Protection of Birds and owner of estates where they kill them in large numbers is one of those subtleties which always escapes this member of the urban lower orders.'

On 6 November the Crown Prosecution Service said that they couldn't take the case any further as no bodies had been found. The *Guardian* reported that:

> *As a result, there was no forensic or ballistic evidence to study. And since all three suspects denied any knowledge of the incident, and there was no eyewitness testimony of who had fired the fatal shots, the case was closed.*

and:

> *Marcus O'Lone, the Queen's estate manager, claimed the failure to find the birds' bodies suggested they had never actually been shot. 'I just can't believe it will have happened' he told the Guardian last week. 'At the end of the day these are allegations. The police are investigating it and we will have to await the results of their inquiries.'*
>
> *But yesterday, the CPS said it had no doubt the birds had been shot. And, it revealed, the prince and his companions were also questioned over the illegal use of lead shot over the nature reserve - an inquiry hampered by the removal of the ducks which had been shot.*

The RSPB commented widely on the result of the investigation as well as contributing a prominent 'Comment is free' article on the *Guardian* website. This article, written by me started:

> *The lack of evidence to justify a prosecution over the gunning down of two hen harriers on the Sandringham estate is disappointing, but not surprising. Wildlife crimes are difficult to prove: rarely are there witnesses, and the victims' families cannot report their loss.* and went on *Increasingly we hear reports of armed gangs visiting winter roosts of hen harriers away from their moorland breeding*

sites in order to shoot several birds in a day. There is one such harrier roost near to Sandringham and this is where the birds were shot on the evening of October 24. Whether this was deliberate targeting of this species we will never know. And it seems we will never know who was responsible either.

and ended:

Although the death of two hen harriers remains unexplained in terms of who was responsible, we hope this will alert the public to the unacceptable practices of a relatively small number of people in the countryside and thus increase the chance that the police will be given the resources and information that will lead to future such incidents being punished by the courts. Our natural heritage deserves to be protected.

By 11 November the *Daily Telegraph* put a different gloss on affairs under a headline 'Dirty tricks claim in Harry "shooting"'. The *Telegraph* said:

Prince Harry is the victim of a 'dirty tricks' campaign over the alleged killing of two birds of prey at Sandringham, his friends fear. Two rare hen harriers 'killed' while hovering over the Queen's Norfolk estate were not even shot at, those close to the Prince now suspect.

Friends of the Prince, 23, are angry that the Royal Family has been dragged into the affair despite no firm evidence that any bird of prey was illegally killed – and Harry could be the victim of a campaign by the anti-shooting lobby.

The Sunday Telegraph has learnt that senior staff at Sandringham - the Queen's country house in Norfolk which has a 600 acre estate - have carried out their own investigation into the 'shooting' with the help of the 'suspects', Prince Harry, his friend William van Cutsem, 28, and David Clarke, a gamekeeper.

> *The inquiry has concluded that there was probably no such shooting and that the supposed 'eye witnesses' were, at best, mistaken over their claims. The alleged incident happened just before 6.30pm when it was nearly dark. No birds' carcasses were found, even though the police were quickly on the scene. It is understood that only one of the alleged three witnesses - a member of staff from Natural England - gave a statement to the police and none of them wants to be identified.*
>
> *One friend of the Prince said: 'Harry is an experienced shot and knows the difference between duck and harrier hawks. He simply didn't shoot the birds. This story was based on, at best, thin evidence. I wouldn't be surprised if someone had an agenda here because they were certainly very eager to stitch the guy [Prince Harry] up.'*

The RSPB was quoted widely in the media, and we did many TV and radio interviews on the subject over this period. Our line was that killing hen harriers is illegal, too much of it goes on and that if anyone was found to be responsible for such an illegal act then it was a serious matter. We were criticised by some of our members who felt we had been silent on the matter – a difficult view for us to understand since our website had featured the incident and we had talked to the media very openly. But we had also been criticised by shooting interests for whipping things up and unfairly stoking the fires of controversy. It is difficult to reconcile both views – except that different people sample the media in different ways and we all see and hear things through our own filters.

It was some time later that we had to decide what, if anything, we should say about the matter in *BIRDS* magazine. The next *BIRDS* magazine to appear after the Sandringham incident was in January 2008, some time after the events had receded into memory. But wouldn't it be odd not to mention this very high profile incident which occurred on the land of our Patron, Her Majesty the Queen?

In the end, Graham Wynne and I wrote an RSPB View piece which led on hen harrier persecution under the headline of 'An old madness' before moving on to the 'new madness' of biofuels. The piece contrasted the almost Victorian continuation of the horrors of raptor persecution with the new ways that we were finding to harm nature through rainforest-trashing biofuels. I've always thought that the juxtaposition of these two subjects showed the breadth of the RSPB's current work very well – still caring about individual acts of unkindness to birds but also tackling the big issues of the day

In that article Graham said: 'The shooting of two hen harriers at Sandringham in October and the poisoning of a golden eagle in southern Scotland last summer were despicable acts and should be sources of shame for those involved' and that, 'The RSPB will remain steadfast in working with the police and other authorities to combat wildlife crime', and 'A Victorian approach to protected species is simply unacceptable in the 21st century.'

Well, it was a while ago and so we may never know what actually happened that evening. All the doubts concerning the original description of the event seem to come from the Sandringham Estate and from shooting interests. Perhaps on the other side conservationists were too eager to believe the worst interpretation of the reports. However, I have spoken to several people involved in the case and none has given me the slightest indication that they have any doubts that two hen harriers *were* killed illegally that evening in late October 2007. And, note, the Crown Prosecution Service did not express any doubts either about the integrity of the witnesses – they only said that in the absence of any eyewitness identification of the perpetrators or forensic evidence there was no case to take forward. My own view is that someone certainly shot two harriers and I really don't know who it was.

Towards the end of October my mind always turns towards those hen harriers shot at Sandringham. I did consider the possibility that we should stir things up on hen harriers on the anniversary of those events every year. Maybe 24 October should be known by birders as 'Harrier Day'?

Sparrowhawks

Sparrowhawks are agile raptors which feeds on small birds. Not for nothing, when asked by Queen Victoria how to rid the Great Exhibition building of house sparrows did the Duke of Wellington merely opine 'Sparrowhawks, Ma'am.'

Sparrowhawks are similar to cheetahs – both are quick, agile, short-pursuit predators that chase down their prey and then eat it. Neither will ever turn vegetarian and both are natural born killers.

For those who were around in the 1970s, or earlier, the return of the sparrowhawk is as impressive a recovery as that of the reintroduced red kite. A day birdwatching in my youth might or might not have provided a sighting of this bird, but I lived in the west of the country, where the decline of sparrowhawks, due to eggshell thinning as a result of agricultural pesticides in their food chain, was much less than in the more arable east.

It was the banning of pesticides such as DDT which allowed the sparrowhawk to spread back to the east of England throughout the 1970s, 1980s and 1990s. Much is made by some people of the large population increases of raptors such as buzzards, kites and sparrowhawks, but these are simply recoveries from the impacts of persecution (some of it carried out many decades ago), pesticides and, in the case of the buzzard, probably myxomatosis too. Rather than chunter on about these unnaturally high populations we should simply be relieved that they have returned.

The sparrowhawk is at one particular end of the raptor conservation spectrum and those who are simply prejudiced against birds of prey, or who deplore their impacts on shooting interests, have spotted this peculiarity. It would, probably, be possible to cull sparrowhawks every year, in just the ways that crows are culled, and not drive the species to extinction. Sparrowhawks are sufficiently short-lived and highly productive that one could, if one wanted, allow sparrowhawk culling in such a way that population levels would be reduced without affecting the range of the species at all.

Of course I am not, in any way, suggesting the authorisation of this sort of culling.

The sparrowhawk was the last bird of prey to be given full legal protection in the UK, in 1961 rather than 1954. So it was the last in – and some, I suspect, wish it were the first out. Sparrowhawks certainly do include grey partridges in their varied diet but partridges are towards the larger end of the prey they can take and their penchant for woodpigeons and other species should make them the farmers' friend. But most of their diet comprises smaller passerine birds and this means that they are the raptor most likely to be seen in peoples' gardens, perhaps taking their favourite great tits, or even the occasionally robin, from the bird table. This visibility makes the sparrowhawk a good candidate for those who wish to blacken the name of all raptors.

There is no sense in disliking some species in the natural world while liking others but we all do it to some extent. We ascribe, to varying degrees, human values to those species, so that sparrowhawks can seem cruel if they catch a blackbird in their sharp talons and then start plucking and eating the struggling bird on your lawn while you watch. If you see such an event whilst eating your Sunday roast, dining on a mammal which was carefully transported to an abattoir, stunned and then humanely despatched (you expect) then the danger is that the sparrowhawk may appear unnecessarily barbarous and a morally bad creature. The argument does not hold up very well intellectually but the sentiment may lurk in the backs of our minds, and I do wonder why cheetahs never seem to get this grief as they commit similar crimes on the cutest of baby antelopes.

But those who really wish to denigrate the sparrowhawk choose a different route – a more complicated route in reality, but one where simple arguments get you quite a long way with quite a lot of people. That is to suggest that sparrowhawks have caused the declines in farmland birds that are such a strong theme of UK bird trends.

Now this is not a completely daft idea, and I wouldn't be surprised if sparrowhawks do suppress the numbers of some species a bit. It's useful to think about how one would feel about this if it were true. This is a very different case from the impact of hen harriers on red grouse that have been produced at artificially enormous densities in order for some of them to be shot by paying sportspeople. If sparrowhawks are contributing to the massive countrywide declines in birds such as skylarks, tree sparrows and turtle doves then that prompts all sorts of new ideas. Let's not blame farmers! Let's cull sparrowhawks – and while we're at it, buzzards too! Let's introduce payments to subsidise the work of gamekeepers! There are a host of vested interests who would sign up, *en masse*, to this agenda.

But wait! Another way of thinking about this is that if sparrowhawks depress the population levels of their prey, then so be it. That's fine by me. It has presumably been the way of the world for hundreds of thousands of years and sparrowhawks have not yet been responsible for the extinction of any species. Let's just let them get on with it as they always have done. That would be my personal view but you can see that any evidence, anywhere, that sparrowhawks have caused declines in farmland birds, garden birds or woodland birds could be presented as damning. So let's look at the evidence.

The most direct test of whether sparrowhawks depress the population levels of their prey would be to carry out an experiment whereby sparrowhawks were removed from large areas of the countryside and then see whether the numbers of other species increased. It might be difficult to do this in the countryside now but 'luckily' for us it was done over a period of decades during which sparrowhawks did disappear because of DDT. At the time, birdwatchers were not all talking about the massive explosion in skylark numbers, or that great tits were charging through our woods in huge flocks because of the relaxation of sparrowhawk predation pressure. It wasn't an experiment as such. It wasn't carefully controlled and you could nit-pick it to bits. But the fact that a near absence of sparrowhawks did not set tongues wagging about an increase in small bird numbers is quite telling. It suggests that if sparrowhawks disappeared again we wouldn't see much difference in the numbers of farmland passerines.

There are a couple of other detailed studies which show essentially the same thing. At Bookham Common in Surrey, the populations of passerine species were recorded in the presence and absence of sparrowhawks over some decades. Did the return of the sparrowhawk lead to noticeable declines in this wood? No, it didn't.

Similarly, in Wytham Wood outside Oxford, where I once wandered around looking for radioactively marked sunflower seeds, a study by Tim Geer showed that sparrowhawks ate large numbers of young great tits, as well as some adults of course. In some years up to a third of the young produced by great tits in the woods went into the crops of young sparrowhawks, but this did not cause a decline in the great tit population despite this enormous harvest. And great tits did not boom in the years when sparrowhawks were absent, nor crash in the years when they were present. Great tits are common sparrowhawk prey items and their populations have held their own, and indeed increased as sparrowhawks recovered in numbers and distribution in the years following the banning of DDT.

Two very detailed studies have looked at the whole BTO dataset on population trends for a wide variety of species going back to the 1970s. Does analysis of this dataset provide any strong evidence for believing that sparrowhawks are to blame for song bird declines? The first study to look at this issue was commissioned by the RSPB and showed that we were taking it seriously. I was Head of Conservation Science at the time and remember wondering whether it was a wise or foolish move, but the scientist in me said that it was always better to know than not to know.

I wavered slightly in this belief when, towards the end of the study, the preliminary results looked rather worrying. The RSPB's detractors may not believe it when I say that we thought hard about what those results would mean for us, but never considered suppressing or ignoring them. However, a few days later the scientists came back, slightly sheepishly, and admitted a computing error which, when corrected, produced results that any sparrowhawk or sparrowhawk-lover could easily live with. The study by Dave Thompson, Rhys Green, Richard Gregory and Stephen Baillie was published in the prestigious journal *Proceedings of the Royal Society B* in 1998 and the paper's title sums up its results nicely: 'The widespread declines in songbirds in rural Britain do not correlate with the spread of their avian predators.'

The second study, essentially an enlarged repeat using the same dataset (but a dataset that had grown in the interim with the addition of more recent data) took a broader view and looked at a larger number of predators. It was funded by Songbird Survival (of whom more later) and carried out by BTO scientists with the help of academics from the University of St Andrews and the Game and Wildlife Conservation Trust. Its results were published in spring 2010 and, as before, if it had been a court of law sitting in judgement on the sparrowhawk the defendant would have walked free. I can't improve much on the summary that I wrote in my RSPB blog at the time:

> *...of the 29 species, the results for only seven prey species are statistically significantly correlated with sparrowhawk numbers – some positively and some negatively. The seven prey species are: lapwing, yellow wagtail, robin, nuthatch, tree sparrow, bullfinch and reed bunting. The two most statistically significant results are that sparrowhawks depress tree sparrow numbers and increase yellow wagtail numbers. I find it easier to believe the former than I do the latter – but who knows? And although I can imagine that sparrowhawks might depress both reed bunting and bullfinch numbers I find it more difficult to believe that they increase robin numbers or decrease nuthatch numbers. And a whole host of regular sparrowhawk prey species (blue tit, great tit, blackbird, song thrush, starling and house sparrow) are unaffected, it seems, by sparrowhawk numbers.*
>
> *This analysis could have come out very differently – it could have shown clear across-the-board negative correlations between*

predators and their prey populations. Or it could have shown just a small number of clear negative correlations for some predators and some prey which were biologically realistic. Neither happened. So let's stop fingering predators as the cause of widespread songbird declines and look to the real causes – unsustainable agriculture, pollution, climate change and the overall pressure of our species on the natural world.

But let's also wonder what we would have done if this analysis had shown that predators reduce the populations of their prey. I think the answer is not much! If that's what happens then let's get the habitats right – as we have at Hope Farm – so that predators and their prey can coexist and let's have many more of both!

Maybe in a future decade someone else will have a poke at this dataset with a different, more powerful, analysis, and maybe we'll have to accept that sparrowhawks do depress their prey populations somewhat. But from where we stand now it's difficult to give any credence to the often peddled view that sparrowhawks are 'the reason' for farmland bird declines.

I think that most of the bad press for the sparrowhawk is generated by that part of the shooting lobby that is waging a public relations war against birds of prey in general, aided at times by a few grumpy farmers who see an advantage in blaming nature for what farming has done to farmland birds. And there will be some people who feel uneasy about predators such as sparrowhawks, seeing them as bestial 'killing machines'– although I really can't understand why they don't hate blackbirds for their equally dastardly treatment of earthworms.

The overall picture

Whether our motives are well-informed or misinformed, whether they are selfish or public spirited but misguided, or whether they are based on some science or purely emotional – we are not, as a nation, very nice to our birds of prey.

In the RSPB annual review of crimes against birds, which includes egg collecting, illegal imports of protected species and a range of other subjects, accounts of people being nasty to birds of prey are by far the most numerous.

The 2010 *Bird Crime* report contains a litany of offences against birds of prey: the shooting of migrating ospreys; the setting of spring traps in peregrine eyries; the poisoning of red kites; attempts to smuggle peregrine eggs out of the country; the finding of three poisoned golden eagles on a single estate in Sutherland; and a hen harrier caught in a leg trap on another Scottish estate.

In 2010 there were 227 incidents involving the destruction of birds of prey, some involving many birds. While this was a decrease on the previous year, it was about average for the last few years – things are bad and not really getting any better.

The reported incidents are just the tip of an iceberg of illegal activity about which we should be ashamed as a nation, and which we should bear in mind when we criticise all those 'nasty foreigners' for killing birds on Malta or Cyprus. Here in the UK raptor-killing is not the preserve of the ordinary working class man in the street but of the landed gentry (aided by a few pigeon-racers who target peregrines).

While many, but not all, raptor species are recovering in numbers their recovery is still held back, slowed down and prejudiced by illegal activities. And, as we have seen, there are still large areas of the country which lack any noticeable populations of birds of prey because of this.

A great deal of the illegal activity directed towards birds of prey is carried out by game-shooting interests. In the 2009 *Bird Crime* report, the RSPB published an analysis of the 141 people who were convicted (not charged, or suspected, but actually convicted) of crimes against birds of prey in the 20-year period 1990–2009 and found that 95 of them were gamekeepers. Gamekeeping is sometimes called a profession, usually by gamekeepers, but if it really were a profession, like medicine or the law, then an awful lot of its current members would be in fear of being struck off and banned from their work for performing it illegally.

I suspect that many gamekeepers live in fear a lot of the time. They must fear the wrath of their employer if the game bags are not high enough and he blames it on a bird of prey seen flying around the estate. They must be in fear on those early spring mornings when distributing poison carcasses or staking out a harrier nest to kill the female. They are damned if they do and damned if they don't, in so many ways. That is why the law in Scotland, but not yet in England, acts against the landowner as well as his gamekeeper if wildlife laws are broken. There is currently an e-petition calling on the Westminster government to introduce vicarious liability into this part of the legal protection for birds of prey, so that law-breaking landowners share more of that fear with their law-breaking employees – please sign it.

The next target – the buzzard

Demonising of particular species comes and goes. If you have a long enough memory then you will remember tawny owls being the *bête noir* of the pheasant rearing industry back in the distant past. Herons and otters have had their moments and magpies and crows are perennial favourites. Foxes and badgers, harriers and hawks, cormorants and pine martens all get their day in the spotlight. My prediction is that the buzzard is next in line but before I tell you why, it is interesting just to note that while golden eagles have attracted the attention of upland sheep farmers, kites, ospreys, hobbies, kestrels and merlins rarely come in for any such stick. I suspect that this is because they are either too small to bother game interests, are seen as being too popular with the public to dare to try to besmirch their names, or they are just being kept in reserve for later.

The next species to be tarred and feathered may well be the buzzard. Buzzards are, I hope you will allow me to say, slightly dull birds. They are perfect for demonising because they are big (and therefore must be vicious killers), obvious (and therefore people notice them), increasing in numbers back to their previous geographical range (and therefore can be portrayed as undergoing a population explosion of phenomenal and unnatural extent), and eat a wide variety of prey species (being generalist predators) and so can be photographed sitting on top of a pheasant or rare bird to make whatever point you wish to make.

Songbird Survival

Songbird Survival is a charity whose strap line is 'Saving songbirds with science' which will seem very reasonable to those who stumble across them and feel they might be best friends with the RSPB. But many who know them a little better would regard this as a complete travesty of their actual ambitions. Songbird Survival did sponsor the most recent wide-ranging study of the impacts (or lack of them) of predators on their prey species which exonerated the sparrowhawk from causing severe and widespread declines in farmland birds. Since then their rhetoric and website have become more reasonable on the subject but they can still be encountered at country shows claiming that predation is a major cause of nature conservation concern – as they are entirely entitled to do.

It is interesting to see who the supporters of Songbird Survival are, where they come from, and what their backgrounds are.

The Chair of Songbird Survival is Viscount Coke from Holkham Hall, which is situated on the north Norfolk coast a short drive from the Queen's Sandringham Estate. Lord Coke's father, the Earl of Leicester is quite outspoken on the subject of birds of prey. In *Country Illustrated Magazine* in October 2006 he says: 'I do not advocate the wholesale slaughter of birds of prey – there is room for everything. I enjoy watching harriers and other birds of prey.' And goes on to say: '...but action must be taken when one species comes to have such an impact on another species.' without it being clear about which predator and which prey species he is speaking. If his Lordship enjoys watching buzzards as well as harriers, then he had one fewer to enjoy when one was found dead, shot, at Holkham in September 2009 when the estate matched the RSPB's offer of £500 reward for information leading to the conviction of the criminal concerned. No prosecution has been brought.

In that same *Country Illustrated Magazine*, the Earl of Leicester is further quoted as follows:

> *Predation is a growing problem, especially avian predation. We need young grey partridges on the ground to produce a viable stock, let alone a shootable surplus. When brood sizes are limited due to lack of insects for chick feed and then those chicks are relentlessly pursued by birds of prey, we seem to be fighting a losing battle.*

That'll be the battle to shoot them then? In which the raptors are relentless while the hunters, presumably, are kindly? I see.

Holkham Estate is a large spread – 25,000 acres – and its team of gamekeepers distributes 200 tonnes of wheat through feed hoppers for game birds which must also feed a few finches and buntings. But, according to the October 2006 issue of *Country Illustrated Magazine,* those keepers also killed, in 2005, 16,296 rabbits, 7,653 rats, 1,209 grey squirrels, 420 stoats, 249 magpies, 221 jackdaws, 179 carrion crows, 188 weasels and 82 foxes. That is quite an annual total and shows the scale of killing that accompanies the more obvious shooting of pheasants and partridges on a shooting estate. I'd like to make a general point here, not related to Holkham specifically. If your working days are routinely filled with this amount of slaughter then there must be a danger that adding the odd bird of prey to the list does not feel like a moral lapse. Perhaps it just feels like another death in a very long list of deaths – and no big deal?

As reported in *Legal Eagle* #25 (the RSPB Investigations Newsletter) in July 2000, Holkham does not have a completely pristine record as far as birds of prey are concerned. In March 2000 a Holkham gamekeeper was found guilty of killing three kestrels.

The former head keeper at Holkham was Simon Lester who is now the head keeper at the Langholm estate, of hen harrier and grouse conflict fame but where diversionary feeding seems to be doing the trick.

The current head keeper at Holkham used to work for the van Cutsem family and Hugh van Cutsem is another trustee of Songbird Survival.

Hugh van Cutsem is a lovely man whose Norfolk estate is very important for stone curlews and other birds. I've always liked Hugh as he comes from a family with a long and deep involvement with horse racing and that has always been a bond between us. It should be said that Hugh operates at the posh end of flat racing – owning horses and a stud – whereas my interests are at the working man's end of things involving the Tattersalls' enclosure at Towcester and Huntingdon – but the bond is real. I visited Hugh's estate on the Thursday of Royal Ascot once with Barbara Young and tipped *Pilsudski* as the winner of one of the big races. Although *Pilsudski* came second (the story of my betting life?) Hugh remembered that and we got on well ever after. Hugh is one of those curious mixes that one finds in charge of shooting estates – keen on shooting (very), keen on wildlife (very), keen on legal predator control (very) and not at all keen on birds of prey. Hugh and his wife played a large part in bringing up the royal princes after the death of Princess Diana and Hugh's son was out on the Sandringham Estate on the evening when two hen harriers were reported to have been shot out of the air in October 2007.

Her Majesty the Queen herself was reported by the *Daily Telegraph* (26 March 2008) to have given money to the organisation Songbird Survival. The money came from the Queen's private income and so is entirely her own business. A spokesman for Songbird Survival quoted in the article said: 'We can confirm that the Queen has made a donation to the charity from the Privy Purse Charitable Trust.' I've always wondered how Songbird Survival came to her Majesty's attention. Was it a mention from Hugh van Cutsem or the Earl of Leicester perhaps? Maybe Lord Peel, ex Chair of the Game and Wildlife Conservation Trust and former grouse moor owner, but now Lord Chamberlain, and therefore the senior official of the Royal Household, might have mentioned them in passing? I'd love to know, but there are probably so many landed gentry with an indisposition towards birds of prey that it would be difficult to narrow down the field to a manageable number.

The frequent links back to game-shooting must, surely, tell us something about the sympathies of Songbird Survival.

Not all the supporters of Songbird Survival are landed gentry involved in gameshooting. The hill farmer John Pugh from Rhayader is quoted on Songbird Survival's website as claiming that he was one of the prime movers in setting up the group. Talking about rural predators – the raptors, crows and magpies, badgers, foxes and squirrels, he is quoted as saying: 'I have no hesitation in saying it's terrorism in the countryside. How else can you explain what is happening?'

I do not know whether Mark Osborne, a game manager from Banbury, is a paid up supporter of Songbird Survival but my guess would be that he might be. Mr Osborne wrote in the *Farmers Weekly* about the 'extraordinary imbalance' of raptors. I've met Mr Osborne a couple of times and he is a well known figure in shooting circles.

Mr Osborne writes in *Farmers Weekly* of the:

> *... massive localised damage being caused by some raptor species'* and *We in the countryside are going to have to accept often significant damage caused by raptors, but equally if this damage is at a level that threatens to wipe out the prey species, or even take their populations to a vulnerable low, then we should be unequivocally pushing for control of raptors in some format. That way we may actually make some progress.*

I think we should be grateful to Mr Osborne for making his views so clear and so public. I've heard similar sentiments from others in private, but Mr Osborne is laudably open in what he thinks about birds of prey. When he writes, 'we in the countryside' I wonder who he means - perhaps he is thinking of the shooting community.

Conclusion

For a nation of animal lovers the British allow an awful lot of birds of prey to be killed each year, to such an extent that the population sizes and geographic ranges of several species are still constrained by this widespread and systematic illegality.

Much of the killing is associated with gamekeepers working for shooting estates – which, I must repeat, is not to say that most gamekeepers are breaking the law. Those who care about wildlife should continue, in my opinion, to publicise these illegal practices and bring them to the public's attention. This may not cure the problem on its own but it is better than turning a blind eye to it.

Is there a solution? I think that it is unfortunate that the good guys in the shooting community have not been able to influence the bad apples within their own barrel. I have met plenty of good apples and plenty of bad ones and I honestly do not know the relative frequency of the two. Maybe the good are just outnumbered, or outgunned, by the bad? Maybe that is why they have made so little progress and why nature conservationists hear so few moderate voices from which to take encouragement. The time for gentlemanly chats on this subject has nearly run out and I would suggest that the time has come to seek allies for a proper attack on the legality of game shooting. The obvious place to start, and in my opinion finish, for it is by far the worst case, is driven grouse shooting. If it were up to me I would launch a campaign to ban driven grouse shooting on 12 August 2013. In fact, that's not a bad idea at all.

We are better than anyone, ain't we? Except for the Eagles, the Eagles are better than us.
Sid Vicious

Chapter 12

Trying to change the world

In the councils of government, we must guard against the acquisition of unwarranted influence, whether sought or unsought, by the military-industrial complex. The potential for the disastrous rise of misplaced power exists and will persist.
Dwight D. Eisenhower

The problems for wildlife in the UK are not usually caused by people who dislike wildlife *per se* but by indifference to, and ignorance of, the impacts of our behaviour on the wildlife around us. The disappearance of skylarks isn't caused by farmers actively disliking those little brown birds, but by a system of farming that neglects the needs of plants, insects and birds in the countryside. And so it is with most environmental problems. It's the way our Society is run – the underlying values that it espouses, often unconsciously, give wildlife a hard time. So, all you need to do to save wildlife is to influence the way that Society works and alter its values – and much of the problem is solved!

That's obviously quite a big job, to say the least, and you could be forgiven for just throwing up your hands in horror and giving up. But experience shows that if you direct your efforts wisely, then you *can* win battles and influence the people who hold power, and by so doing get a better future for wildlife.

This chapter looks at who you have to influence and how you can go about influencing them.

Which public policies to influence and how you need to influence them

Birds are everywhere, all the time – in every habitat throughout the UK and at all times of year. This means that almost anything that happens will influence birds. So we are dealing here not only with wildlife legislation, which is vital in protecting species and their most sensitive sites from direct threats, but with all those other public policies such as housing, transport, energy consumption and production, as well as the major land uses such as forestry, agriculture and fisheries. Government policies and spending on energy, agriculture, forestry, building, transport and fisheries set the rules by which the species in our fields, woodlands and seas must live – and they don't get a vote. It isn't just the 'environment' department that needs to be influenced to give nature a better deal. It's the 'foreign' department, the 'home' department, the 'local government' department, the 'education' department and, fundamentally, the 'money' department – because money talks so loudly.

In October 2010, the new coalition government announced the results of the Comprehensive Spending Review which looked at government spending by government departments. Whichever political party had won the May 2010 general election, there would have been cuts in spending but with a different party in power the size of the cuts, and their allocation across different policy areas, may have been radically different. As it was, probably through a combination of bad luck, some unfortunate tactical choices by Defra (only identifiable in retrospect) and conscious decisions by the incoming government, Defra received just about the biggest *pro rata* cuts of any government department. This is having an impact on the number of civil servants within Defra itself, and also on the funding of Defra agencies such as Natural England and the Environment Agency. Natural England suffered serious job losses to accomplish its targeted 30% cut over four years and those cuts had knock-on impacts on its funding of all the NGOs with which it had partnership projects. If the cuts had been smaller, wildlife would have been less affected.

The new government policy on energy generation and use affects wildlife too. Changes in the proportion of our energy needs to come from renewable and non-renewable sources, and whatever proportion is planned from nuclear power, will influence the number of windfarms in our countryside and the number of proposed estuarine barrages that might affect our estuaries and their wildlife. This wide-ranging policy area is influenced by the Treasury but also by the various government departments dealing with business, as well as the relatively new Department for Energy and Climate Change. There are a lot of different doors on which to knock if an NGO is to influence the wide range of wide-ranging policies that will affect the wildlife around us.

However, while this plethora of government departments whose focus is not on the natural world, and yet whose policies will affect wildlife, is important the most important relationship of all for wildlife NGOs must be with Defra – the government department with responsibility for biodiversity as well as agriculture, etc. And the goal of any wildlife NGO must be to make this a close relationship. Defra must be kept onside because it is the job of Defra Ministers to be on the side of nature, albeit while trying to keep voters happy and also looking out for their own careers at the same time.

Every now and again Defra has a legislative slot in government business that enables a leap forward for wildlife. In recent years the Countryside and Rights of Way Act (which required good management of SSSIs) and the Marine and Coastal Access Act (which gave marine wildlife better protection) were both important legislative advances. In both cases close cooperation between NGOs and Defra achieved cross-government support and enabled their entry to the statute book.

Big changes in legislation and policy come up every few years – reform of the Common Agricultural or Fisheries Policies, fiddling with the statutory agencies that do nature conservation work, new legislation or changes to funding, etc. And in a democracy everyone can try to influence what happens and NGOs like the RSPB play their part. Indeed these occasions are often where disparate NGOs work most closely together.

Decisions that influence the future of wildlife are made at every scale. Globally, the World Trade Organisation's rules make it difficult to discriminate against imports on the grounds that they have been produced in ways that harm wildlife. Nearer to home, your local council may make planning decisions that will affect your local wildlife. Life used to be much simpler but now what happens to wildlife in your locality will be influenced by global climate change decisions (or lack of them), EU Directives concerning wildlife, water use and so on, UK government decisions on funding, and truly local decisions made within your parish.

The process of devolution within the UK means that it is rarely appropriate to talk about 'the' government anymore because UK wildlife is affected by nature conservation policies which may well differ in detail in Belfast, Cardiff, Edinburgh or London. It can be difficult for we English to understand that our government in Westminster sometimes acts for the UK – on subjects such as defence and taxation – and at other times is the English government – on subjects such as nature conservation. Taken together, globalisation and devolution mean that if you are to influence the policies that influence birds and other wildlife then there are an awful lot of people to influence in an awful lot of places.

One thing you can be sure of – nature won't be represented in the rooms where important decisions are made unless people speak up for nature. The skylark and the marsh tit don't get a say, or a vote, and it is only if people speak up for them that their natural world will be considered. You can't expect that civil servants and politicians will necessarily be thinking about nature and so they need a bit of gentle reminding sometimes.

The difference between policy and advocacy

I find it useful to make a distinction between policy and advocacy. Policy is defining and refining what you, or your organisation, think. So the RSPB would have policy, made up of lots of sub-policies on, say, agriculture. It was part of my job to bring that policy together every now and again for approval by the RSPB's Council, which has the final say on the RSPB's policy. Luckily, I had lots of clever experts who would do most of the thinking but it all had to be summarised at particular times, and in ways that Council would agree, which would also give staff the necessary guidance, and freedom, to get on with their jobs too.

That whole process of policy development is vital for an organisation dealing with as many issues as the RSPB. I saw my job as Conservation Director to ensure that we had the policies in place, that they were signed off by Council, and that there weren't too many awkward contradictions between one policy area and another. There's a need for coherence – and it is essential that the bundle of approved policies for any organisation do not conflict with each other, or with the way that the organisation actually works. It's particularly important to walk the walk, if you talk the talk.

Once you have your policy line on a subject then you need to advocate it. The rest of the world has been defining and refining its policy whilst you have been producing yours and now you have to try to make the rest of the world believe that your policy is best. The perfectly crafted policy is no use at all unless it is winning hearts and minds out there in the real world. It's not enough just to be right, logical and clever you also have to be persuasive.

Some of the people who are best at policy development are not the best at persuading others to adopt your policy. Logic and being right are a good place to start but you have to seem relevant as well. Think about how you have been persuaded to do things. Did you always listen to the logic of the argument or did you sometimes do things because you liked someone, owed them a favour, wanted to please them, or because it was easier to do it than cause a fuss by not doing it? Maybe, on occasion, you were made to do something, or scared of the consequences if you didn't. All these motives and emotions come into play when decision makers decide what to do – and it's best that you realise that when trying to persuade them.

That isn't to say that being well-informed and right and logical isn't very important. If your organisation has a reputation for 'knowing its stuff', then you will be listened to. If you present a good case then it certainly makes it easier for a decision maker to justify their going with your proposition rather than someone else's – even though they might have done so as much to prevent your making a fuss as because they were really persuaded by your argument!

So, advocacy depends on knowing your target audience and then deploying your arguments in the best ways possible. The ultimate target audience for any nature conservationist is the government of the day. If only one could have a ten minute chat with a government Minister every day then I'd warrant that the world would be a better place for wildlife. Unfortunately the opportunities for those chats are few and far between and there are too many other people chatting with them when you are not on the scene. So in seeking to influence government policy one has to employ a variety of other devious networks – MPs who are not part of the government, the civil service, the media, and the public (especially the voting public). For the rest of this chapter I'll be talking about how to deal with these different, but overlapping, constituencies.

The Government is made up of Ministers

If one is to affect political outcomes then there are many ways to do it – but it is usually necessary to get to the man or woman at the top. Sometimes that might mean the Prime Minister but more often it means a Minister. They are very busy people and it is quite difficult to get the attention of busy people at the best of times.

The British system presents some pretty impressive barriers to political influence – the Ministers you wish to deal with may lack interest, lack knowledge and be sent to another job or the back-benches before they have gained much of either. But occasionally one encounters a Minister who makes a huge difference very quickly. Within just a few days of David Miliband taking over Defra from Margaret Beckett, the whole thinking of the department changed – I've never seen anything like it before or since. Mr Miliband's arrival signalled an immediate sea-change with a strong emphasis on climate change. Suddenly, that was what civil servants were talking about and it was clear that the impetus came right from the top. This led to the Climate Change Act of 2008 – a ground-breaking piece of legislation establishing binding targets for greenhouse gas emissions for the UK. But it also shifted the department's attention away from wildlife, making progress in this area a bit slower. David Miliband was one of the few politicians to come into Defra on his way up the political ladder, his next job was Foreign Secretary, and he used it to make a big and lasting difference for the environment. I feel that David Miliband, now on the edges of political life thanks to so narrowly losing the Labour leadership election to his brother, could give aspiring Ministers some tips on how to make an immediate impact on arrival in a new department.

With justification, you might think the system of appointing Cabinet positions to be somewhat odd. The incoming head of a government department, exerting huge amounts of power and influence, does not apply for the job, is not interviewed to judge their suitability for it, or even usually asked about their interest in it! There are few big companies that would do it like this.

A Prime Minister choosing their Cabinet has a certain number of jobs and a much larger number of potential post-holders. In choosing who does what, the politics demand that certain people must get jobs to reward them, or at least keep them on side. Cabinet posts are ranked such that the great offices of state – Home Secretary, Foreign Secretary and Chancellor – are rated much higher than the others, and definitely higher than the Environment.

Each Government department has a 'top minister', usually called the Secretary of State, who sits in cabinet meetings, and a number of junior Ministers. The second ranking is called the Minister of State and the most junior, the Parliamentary under Secretary. In addition there is someone to represent them all in the House of Lords, who has their own share of the Departmental responsibility. It is a world of SoSs, MoSs and PUSs!

Ministers have two types of Special Advisors – civil service experts and political experts. It's often as important for an NGO to have a good relationship with the Special Advisors as with the Minister.

Government Ministers implement government policy as set out in the political parties' election manifestos. In the run up to each general election, NGOs will try to persuade the different parties to include promises, or at least nods in the right direction, on environmental matters. We have had success in recent years in areas such as environmental education and marine legislation, such that all the parties had committed in some shape or form to do something, which has meant action whoever got into power.

There have been some Ministers, certainly junior Ministers, who have appeared to have very little interest in the natural environment during their term of office. Their brief sojourn in the Environment has been a mere stepping stone on their ways up or down the career ladder. On the whole, and there are some exceptions, those politicians on their way down have been the kindest friends to the natural environment. That's probably because a politician on their way up is looking to keep a clean nose rather than to shine, particularly if they don't see the Environment job as very exciting. Keeping your nose clean leads to timid Ministers who don't make radical decisions. And without radical decisions the *status quo* remains – and the *status quo* is not that great for wildlife.

On the whole, environmentalists should always pray for their next Minister to be somebody with passion, guts and clout who has had a successful career already and can spend their last few years of ministerial power doing good for the environment. Two very different politicians exemplified this for me back in the late 1990s and early 2000s: John Gummer and Michael Meacher. You couldn't find two politicians with much more disparate views on most political issues and yet the years that they spanned were good ones for the natural environment.

John Gummer was Agriculture Secretary in the then Ministry of Agriculture, Fisheries and Food before becoming Environment Secretary in the Department of the Environment. He used to be proud of the fact (and I hope he still is) that he was the Agriculture Secretary who removed the grants which encouraged farmers to get rid of hedgerows and was the Environment Secretary who introduced grants for hedgerows to be planted. A pro-Europe, Anglo-Catholic, with a very good brain and an amazing gift for oratory, Mr Gummer's demise was trailed, it seemed, at every Cabinet reshuffle running up to the end of the Major Government and yet, thankfully, he held onto his job and grew and grew within the role.

Mr Gummer did many good things for the natural environment. In particular, he championed the protection of biodiversity after he and John Major returned from the Earth Summit in Rio de Janeiro in 1992. Mr Gummer made speeches about the moral imperative to conserve life on Earth and once waxed lyrical about why we should all be on the side of the dung beetle. Such a politician sends clear signals to the thousands of civil servants working in his department and for a while biodiversity conservation became the big thing in DoE. Knowing that your Secretary of State is keen on a subject always galvanises the civil service into action and if your boss is talking about dung beetles then it is time to start thinking hard about how to take care of them. During these years, the apparatus to save threatened species across the UK, called the Biodiversity Action Plan, was set up. For a few years, it did quite a lot of good but eventually fell into some disarray through a combination of lost ministerial commitment, civil service neglect and the disruption caused by political devolution which meant that responsibility for any UK nature conservation programme suddenly became shared between four administrations in Westminster, Cardiff, Edinburgh and Belfast.

When the Major government lost the 1997 general election we wished a sad farewell to Mr Gummer and looked forward to working with a new bunch of DoE Ministers, with Joan Ruddock looking a likely candidate to get a biodiversity brief. Changes in departmental structures meant that DoE disappeared to be replaced by a bigger DETR under the leadership of John Prescott, and the biodiversity brief fell to Michael Meacher. There were no whoops of joy at this news. Mr Meacher was an old-Labour politician, who had been a Minister under former Labour Prime Ministers Harold Wilson and James Callaghan, and had shown little previous interest in the environment. There was a strong chance that his would be a dead hand at the tiller after the engaging and engaged Mr Gummer's strong personal interest.

But our fears were proved completely wrong. Michael Meacher was an exciting biodiversity Minister. He scared his civil servants rigid – because he always wanted to do the right thing and was brave enough to be unpopular. Meacher had been a Minister in the previous Labour government, back in the 1970s, and he must have known that he was not marked out for advancement under Tony Blair's New Labour regime. So, again, we had a politician without much incentive to keep his nose clean – and so he didn't. Meacher championed biodiversity conservation, made progress in banning the use of lead ammunition over wetland areas and played a large part in making sure that GM crops were tested for their environmental impacts rather than just being adopted on the say so of the biotechnology companies such as Monsanto.

I remember several meetings where we sat down with Mr Meacher and he asked a group of NGOs about what he should do, listened to our advice and then, in front of his civil servants, agreed with much of our line. As often as not, civil servants sitting behind their Minister looked rather worried about what he had just committed them to do – but that's how it's supposed to work. Ministers decide policy and civil servants carry out their political masters' wishes.

That Meacher followed Gummer at the change of government was in retrospect, if not at the time, a very happy coincidence. Not an entirely golden age for government nature conservation, but certainly one which sparkles quite brightly still from this distance. Two men, each very different in personality and political views, but both principled and both politically adept. The baton of nature conservation passed more effectively between them than it often does between successive Ministers from the same political party. I wonder whether both, knowing that their environment job might well be their last ministerial position (Gummer because it was at the end of the conservative regime and Meacher because of his age, unpopularity with the Prime Minister, Tony Blair, and the need to offer jobs to the ranks of New Labour MPs), woke up in the morning fired with the desire to do good. Did they think that today might be their last day in office, ever, so they had better do something good before it's too late? That's the spirit one needs in every Minister.

All the other MPs

There are 660 MPs in the Westminster parliament and they can't all be in the government. Some were on the losing side in the previous general election, but an awful lot of the winning side are just too inexperienced, too difficult, or too numerous for the limited number of jobs available. MPs play a variety of roles in helping to influence government policies by representing the views of their constituents, sitting on Select Committees to examine policy areas, quizzing Ministers, and writing reports. Very occasionally they get their very own chance to invent some legislation.

The Private Members' Ballot is open to all non-ministerial MPs and occurs at the beginning of the Parliamentary Year. It is a ballot, and if you win the raffle you may put forward a bill of your own to become an Act of Parliament. Each year the top seven or so names drawn in the ballot are likely to have at least one day in the sun as their proposed legislation is debated. Sometimes, government adopts a Private Members' Bill and then it is very likely to become law.

In 2001 John Randall, Conservative MP for Uxbridge, now Deputy Chief Whip, came top in the Private Members Ballot. As a keen birdwatcher John had worked closely with the RSPB for many years and was someone whom the RSPB kept in touch with. I might as easily run into him at the Bird Fair as in the Palace of Westminster. John came to us and asked what might be a good subject for a Bill to help wildlife. We suggested a variety of subjects and John chose to introduce, in July 2001, a Marine Conservation Bill which became the forerunner of the Marine Act. The Bill sought to establish marine SSSIs and received quite a lot of government support so that it progressed a long way through the parliamentary system and received its third reading in March 2002. The Bill failed in the House of Lords due to the tabling of a large number of amendments by a small number of peers, many of them fellow Conservatives and some, I have always suspected, prompted by the ports industry. Notwithstanding its eventual failure to become law, John Randall's Bill progressed further than most Private Members' Bills and certainly helped to add impetus to the Marine Act which received Royal Assent in November 2009.

John Randall's Bill was an unusual opportunity and shows the value of developing links with friendly MPs across all political parties. MPs are all different, with different interests, so there is likely to be a smattering of them who are wildlife enthusiasts, and they will be distributed across all parties and constituencies the length and breadth of the land. Working with friendly MPs is part of trying to influence the government of the day. They will ask questions, write letters and simply have conversations which might help to get your views across.

Select Committees examine the work of departments. There are committees in the Houses of Lords and Commons and they hold enquiries into a whole range of topical and important subjects. When they call for evidence it is an opportunity for NGOs to pitch in and sometimes give evidence in person before the committee. More importantly, Ministers will often be called to give evidence and answer questions. The Environment, Food and Rural Affairs Committee examines Defra's work and the Environmental Audit Committee looks at environmental progress across government – and provides a very useful spur to government's work.

I've given evidence to Select Committees over a dozen times and it is a slightly scary process, since everything is formally recorded for posterity and may sometimes be broadcast live on the Parliament channel. And you can be asked anything the committee wants to ask. I remember once giving evidence to a Lords Committee on organic farming and having to defend magpies and sparrowhawks in my evidence.

The late Gwyneth Dunwoody MP chaired the Transport Committee from 1997 to her death in 2008 and so when colleagues and I gave evidence on the impact of ports and airports on wildlife we knew that the independent veteran MP and ex-Minister knew her stuff. We were once caught out and looked rather foolish as we waxed lyrical about the cargo-stacking facilities in the port of Doha (which if emulated in the UK would reduce the need for physical port expansion). Sadly, when asked where on Earth Doha was we both looked blank and the ports' industry representatives, whose evidence followed ours, were able smugly to inform the Committee that the RSPB must have been referring to Qatar in their evidence. Another time I was heading off to some posh early-evening do after giving evidence in front of Mrs Dunwoody's committee and so arrived in a black tie outfit – the Chair raised a slightly quizzical eye and nodded approvingly before starting to grill me on airport expansion!

As well as Mrs Dunwoody, I have watched a great many other talented parliamentarians asking the right questions of Select Committee witnesses. At the moment the only Green Party MP, Caroline Lucas is a strong performer, as is the Conservative MP for Richmond, Zac Goldsmith. Through their questioning, MPs on Select Committees can embarrass government departments sufficiently for them to think again. The virtue of ministerial responsibility is that no Minister likes to look foolish – if a committee gets a subject between its teeth and shakes it a bit then you can be sure that the Minister will be asking questions back in their Department a few minutes after finishing giving evidence. NGOs and other groups who give evidence can help shape the committees' views so that the work of departments will be examined in the right way.

Influencing the public – start with your members

The RSPB starts with one enormous advantage when talking to politicians – the size of its membership. At 1,076,000, the RSPB membership is considerably more than the combined total of the three major British political parties (Conservative, c. 250,000; Labour, c. 166,000; Liberal Democrat, c. 60,000) and therefore not a constituency that any politician would wish to alienate. Of course, the RSPB's Chief Executive doesn't have a block vote and cannot send out the massed ranks of birdwatchers, nature conservationists and environmentalists to bring down the government – although that is something to aim for in the future, I believe!

But politicians do take notice of our membership. I was speaking at a fringe debate at a Labour Party conference one year, sitting next to Ed Miliband who was then in charge of the Party's manifesto, before he became Secretary of State at DECC and before he became Labour leader. I noticed that he pricked his ears when I mentioned the size of the RSPB's membership – that we had at least 60 members in every one of the UK's 600 plus constituencies and that our aim was to mobilise them at the next general election. Then he put his pen down and listened very carefully when I went on to say that there were about 50 constituencies where the number of RSPB members exceeded the parliamentary majority of the sitting MP.

Similarly, Chris Huhne, then the Liberal Democrat Shadow Secretary of State for Environment, and I had a slightly spiky conversation about the Severn Barrage in his office in Portcullis House across the road from the Palace of Westminster. I couldn't resist telling Mr Huhne that there were twice as many RSPB members in his Eastleigh constituency than the size of his own majority. When he asked whether he could have their addresses I pointed out to him that it worked the other way around – we could ask them to contact him whenever we wanted!

It's sometimes said, often by those who know the RSPB badly and wish it little good, that the RSPB's membership is full of urbanites with little knowledge of the countryside. Generally speaking this is the wrong way around. The RSPB's membership is most strong in those rural constituencies where you might see quite a lot of birds. Again, at party conferences, I sometimes used to point out that John Gummer, the Conservative MP for Suffolk Coastal and now Lord Deben, was the luckiest MP in the House of Commons in having the largest number of RSPB members in his constituency (which also includes Minsmere) and that, not surprisingly perhaps given the 'R' in our name, the RSPB's weakest constituency was that of West Belfast, whose MP was Sinn Fein's Gerry Adams.

But, as a charity, the RSPB must be non party-political even though it can be political. In other words, we must try to influence the disposition of government power, whoever is in power, for the benefit of nature and to meet our charitable goals, but cannot support a particular political party, come what may. I'll return to how and whether nature conservation is a party political issue a bit later, but it certainly is a political one.

For a long time the RSPB magazine, *BIRDS*, has carried an opinion piece at the front under the Chief Executive's name. It's well worth going back and looking at those under Ian Prestt's name, written in the 1980s, to see how the nature conservation scene has and hasn't changed, and how the RSPB is dealing with new manifestations of old problems. The opinion pieces that later appeared under Graham Wynne's name were all drafted by me before Graham and I finished them off together. RSPB View is the last bit of *BIRDS* to be written, although it is written over a month before it begins to fall through letterboxes all over the country over a several week period. The task in writing the RSPB View is not just to address *BIRDS* two million readers (though I'm sure that only a small proportion read this part of the magazine) but to imagine that the Environment Secretary is reading it too. The members have occasionally reacted badly to a phrase or a subject – with a million members keeping everyone happy is impossible – but we also knew that politicians have occasionally winced at the mild criticism often injected into the column. Although the tone always aimed for constructive criticism, it could be critical nonetheless.

The RSPB is a powerful insider-track organisation – by which I mean that a lot of its influence is deployed out of the limelight. We speak behind closed doors, in earnest conversations with decision-makers, in responses to government consultations, and in evidence at Select Committees, rather than by protesting in the streets or otherwise shouting our mouths off. Well-informed argument deployed with clarity and passion, and without fear, would be a nice way to describe our approach, but increasingly, and rightly, we have adopted a more campaigning approach to our work. If you have a million members and more, you need, I think, to demonstrate to them what you are saying – which means saying it clearly in public as well as in the corridors of power – and to involve them in saying it too.

As a new Conservation Director I remember Phil Rothwell coming to me and suggesting that we reinvigorate the campaign letter writers – a group of RSPB members whom we could ask to write to decision makers, often MPs, to put pressure on them. I immediately agreed and over the years a number of people have built up the RSPB's ability to ask thousands of people to email or write to decision-makers at the drop of a hat. Ben Stafford, Sally Webber, Martin Harper, but particularly and recently Steven Roddy, have done a great job on this. But so have all those RSPB members who have written to their MPs or other politicians to express their views. Just showing that you care means a lot. Politicians crave popularity – their jobs depend on it in a democracy – and pleasing the voters is essential to them. If your voters ask for something then you often want to give it to them. So let's get asking!

While size really isn't everything, it does help with campaigning. In 2010, the 210,567 bird of prey pledges followed by over 360,000 signatures on the RSPB's Letter to the Future campaign brought our messages home to politicians in a very effective way. Letter to the Future, campaigning to make government cuts less harsh on nature, was backed up by email campaigns to Cabinet Ministers and even weeks of placards and banners in a few key constituencies. We know that this campaign was welcomed by many RSPB members, and that it was noticed by the politicians it targeted, even though some RSPB members were irritated by the approach. We got feedback on ruffled feathers from some Ministers and a more relaxed view from others. Such feedback is useful – it is sometimes the only way that you really know you have been noticed! After the results of the Comprehensive Spending Review cuts were announced in October 2010, we heard a lot of feedback from reliable sources that the RSPB's campaigning was very influential in adjusting where the axe fell.

The media – everyone reads and listens

If you have a meeting with a government Minister you can say what you like to them – within reason and the bounds of politeness – and they can say what they like back to you. But no-one else is listening. If you say what you would like the government to do on the Radio 4 *Today* programme then over a million people have heard it (fewer will have listened) and if that is backed up by occasional mentions on other radio stations and TV, and a smattering of pithy quotes in the day's newspaper then you can be sure that your views have been noticed by the government too. And as a membership organisation, dependent on the voluntary contributions of over a million people, using the media to talk to the world always seemed a good idea to me.

As the RSPB's Conservation Director I worked as closely with the media team as I did with my own staff – in many cases rather closer. And I got to know many of the environmental correspondents of the newspapers and their counterparts in the broadcast media quite well. Over the years, I grew to have a feel for what stories would work with the media and which ones would not. Sometimes one gets so close to an issue that one begins to feel it is the most interesting and important thing on Earth – but the rest of the world may not share your view. There were plenty of times when I was disappointed that our great news story bombed completely and others when I was delighted that something was picked up by all outlets.

One of the friends I made in the media is Mike McCarthy, the environment editor of the *Independent* newspaper. Mike writes like an angel and is one of the few current journalists with a very strong feel for nature who wants to write about wildlife. He taught me that there are two ways that things get into a newspaper – if it's interesting and if it's important. If a story is both then you've got it made but otherwise 'interesting' usually beats 'important'.

Journalists, like politicians, are not generally respected by the wider population but, on the whole, I've found that they are pretty good people to work with. You have to understand their world to get the best out of them. They have deadlines which would scare anyone else rigid – newspapers have print deadlines and broadcast media are planned to the second. What they want from you is a clear message, quickly. If you are talking on the phone to a print journalist then you need to get the facts across quickly to someone who isn't an idiot but who certainly hasn't been living and breathing the intricacies of the planning system, the Red Data List, or the ups and downs of the hen harrier like you have been.

Print media operate to print deadlines whether they be monthly magazines, Sundays or daily newspapers. On a daily newspaper the key time to get your story in is probably during the early afternoon. Journalists don't seem to surface until mid-morning and then they go out to lunch! That period between 2 p.m. and 5 p.m. has traditionally been when your story gets into the newspaper. Rarely does an environment story make the front page and rarely will your story kick something else out of the paper at the last moment. Pictures can sell a story – particularly a wildlife story – so make sure that you have some great images.

Generally speaking the daily newspapers cover wildlife stories quite well but it is rare these days for a story to appear in all the papers on the same day and so it can sometimes feel like quite a lot of effort to phone around lots of newspapers only for it to appear in just a couple of papers at the end of it.

As well as the incomparable Mike McCarthy, other newspaper journalists of note include Charles Clover, Geoffrey Lean and Stuart Winter.

Radio and TV are worth targeting, particularly if you can organise an interview and more particularly if you are interviewed live. The great thing about live broadcasts is that you can't be taken out of context and you can't be edited. What you say is what goes out – and of course that's what makes it scary too – if you cock up then the world heard it there first!

I've already mentioned the Today programme on BBC Radio 4. This was always a target for the RSPB as its listeners are the type of people who are RSPB members and decision-makers certainly listen to the mixture of politics and current affairs that the programme serves up at the beginning of the day. I've been interviewed on the programme many times, around 50 I think over the years, on subjects as diverse as hen harriers, the Big Garden Birdwatch, non-native species and the RSPB's investments.

Sometimes these interviews have been done in the studio sitting across the table from John Humphrys, James Naughtie, Edward Sturton or whoever. But more often it was from a tiny studio wearing headphones at the RSPB Headquarters or sometimes at the BBC studios at Milbank or even over the telephone from home. It's not too difficult being interviewed over a phone line but if a three-way conversation is intended, where you are most likely up against another person with opposing views, then it is actually quite difficult to know when you can and can't talk.

I've spent quite a few sleepless nights thinking about exactly how to put across a point only to be phoned early the next morning to be told that we had been bumped off the programme because of world events. And I've had early morning orders that I be picked up by a car and taken to a field in the middle of nowhere before dawn to give outside 'atmosphere' to a story, often with Tom Fielden or Sarah Mukherjee, whilst talking about garden birds, declines of woodland species or coastal defences.

Sometimes these interviews went well, and sometimes less well, but I could always be sure that friends, relatives and colleagues would come up and say 'I heard you on the radio this morning' and that made me believe that my message had probably got across to Ministers and government departments too. Although I often found that if I ever quizzed someone on what they thought I had said then they were sufficiently hazy to lessen my confidence in how well the message had been transferred!

The rise of social media

The internet means that there are many more 'media' outlets than there used to be. As well as the BBC and national newspapers having their own internet sites, every man, and a few of their dogs, can write and put their own views to an apathetic world. Social media sites such as Facebook and Twitter allow you to interact with lots of people, get their views and perspectives, and give them your own.

This is all happening at a time when coverage of wildlife issues is falling, in the print media at least. Fewer journalists are allocated to rural and environmental issues and they seem to me to getting less, and lower quality, coverage. The internet allows anyone with an opinion, such as a campaigning NGO, to promote their position in their own words, at their own time, without having to persuade an environment correspondent to persuade their editor to include a piece in the paper.

Although the rise of the blogosphere allows any old nutter to write any old rubbish, that was somewhat true of the printed press too. I haven't read every book ever written, not even every bird book, and I don't intend to. Nor do I watch every TV programme or listen to every radio broadcast. But I do sample them all and have my favourites, as well as a longer list of publications and programmes which I dip in and out of. The proliferation of web logs (blogs), with the collected thoughts of all and sundry, can be dealt with in the same way. If you have interesting things to say and can say them in interesting ways, then there has never been a better time to get your views across. Having the RSPB web site as a blogging platform and the post of Conservation Director as one's professional identity, gives one a good head start over all the other nutters.

Between 2009 and 2011 I wrote a daily blog on the conservation issues that concerned the RSPB at the time. My blog gained a following of fellow nature conservationists, RSPB members, civil servants, journalists and politicians. It's easy to get politicians' attention – just write about them! Using a blog to highlight what the government is doing, what it should do, what Ministers have done well and what you believe they have done badly means that you get their attention pretty quickly. Then, as a campaigner, the trick is to use the fact that the Minister may get a copy of your blog on his desk first thing in the morning, along with the summary of the papers and transcripts from the Today programme, to influence him. I used to think it would be great if one could have a five minute chat with an Environment Minister every morning – what an opportunity that would be. Well a blog isn't quite that good – but it's as close as we are going to get in the real world.

You know you've got the attention of government departments when they complain about what you have written or, more occasionally, thank you for it. Provided that you are consistent with the views of your organisation, and with the views that you put to Ministers and civil servants face to face, then having your say through a blog is now an accepted part of the means of influence.

After I left the RSPB, I published a collection of my blogs in a book called *Blogging for Nature* and was delighted that Hilary Benn, the former Environment Secretary, wrote a foreword to the book which I could not have written better myself:

> *If I was ever in any doubt about what people at The Lodge thought I should do next, all I had to do was bring up Mark's blog – if my advisors hadn't already put his latest entry into my red box that night!*
>
> *Being a Minister means listening to every piece of sensible advice you can get your hands on to help you make the best possible decision as well as facing criticism from all sides. But when you're making the big decisions you trust people like Mark to tell it like they see it. Mark's blog was in parts honest, irascible, demanding, exciting and always readable.*

That's a pretty good explanation of why NGOs who know what they are talking about should use these new means of influence much better than they do at the moment. I can see the role of blogs growing in time as that of the print media continues to fall away.

Ministers do nothing – the Civil Service does a lot

Not many people speak up for civil servants but I will. I have met many very good, clever, hard-working civil servants and their job is a difficult one which I couldn't do. I have too many opinions and a civil servant has to be both civil and a servant to government policy. The Civil Service is the bureaucracy that implements government policy and that means that the day after a change of government the Civil Service has a different job to do and needs to start doing it – and that requires a detachment I simply don't possess.

Each Government department contains a hierarchy of civil servants from Grade 1 at the top – the Permanent Secretary for each department (the top Mandarin!) – through Grades 2, 3, 5 and 7 (Grades 4 and 6 seeming to have disappeared, presumably in the course of some reorganisation or another). The further down the scale you go, the more the jobs have to do with administration. The further up you travel, the more they have to do with policy development and elaboration until the top jobs become very political – even if with a small 'p', rather than Party Political (with big 'Ps').

A high-flyer in the Civil Service will be moved on to another job every few years, most likely in a different department, so as to gain a range of experience in the various issues of government. The unspoken assumption is that a good civil servant will have a razor sharp brain that can be turned to any problem at any time, rather than much knowledge of any particular subject. All this means that many of the civil servants one meets do not know very much about nature and sometimes aren't around long enough to learn about it either. Just as with Ministers, the NGO aiming to change government thinking through Civil Service contacts has an ever-changing audience to convince.

Which political party is best for wildlife?

The personal qualities of Ministers make a huge difference to their effectiveness and how much difference they make to the world, regardless of their political affiliations, and the Gummer/Meacher period demonstrates that it's not all down to political party. Most of my interactions with politicians have been during a Labour government but include the end of the last Conservative administration and the beginning of the current Conservative/Liberal Democrat coalition. So is one political party 'better' at the environment than the others?

Now is probably the time to state that my own political convictions lie to the Left rather than the Right. I've been a member of the Fabians on and off (mostly off, but sometimes on) and joined the Labour Party after their defeat in the 2010 General Election – I jumped onto the holed ship rather than off it!

So, rather bizarrely perhaps, I've often found Tories individually to be a lot more sympathetic to the natural environment than Labour politicians. There are far more Tories than Labour politicians whom one can imagine having a pair of wellies in the back of the car, and a pair of binoculars too, although there may be a shotgun lying beside them. For example, the former Shadow Environment Secretary, Peter Ainsworth, is a good example of a civilised Tory who 'gets' the natural environment with a passion that I recognise and share, whereas the former Environment Secretary David Miliband, despite being an immensely impressive thinker, and fully committed to tackling climate change, comes across as an urban intellectual with little feel for the natural world. Such people impress me but I don't feel as though their environmental outlook is similar to my own.

It's too simplistic to say that the Tories get the countryside because they own it, and get nature because they shoot it, whereas Lefties stick to their towns and are strangely in awe of farmers because they've never met one before and don't understand a word they say. Yes, it's too simplistic, but contains a grain of truth.

But let's go to the philosophical differences between Left and Right – surely there is a difference between the parties?

In incredibly simple terms, the strength of a right wing environmentalist is that she will emphasise the responsibility and power of the individual to do good, be a good steward and protect the environment without regard to, or interference from, the State. But, to my mind these strengths fall short if there aren't enough green Tories out there.

The strength of a left wing environmentalist is that she will use State interventions to ensure that the common good of a protected environment is maintained, despite it being in the short-term interests of some individuals to destroy it. But this model falls down if the State does nothing, or does the wrong thing.

The Market is a very inefficient way to deliver environmental progress because environmental goods aren't marketed. If you are a farmer with lots of skylarks on your farm then you can't easily make money out of them – but you may be able to make money out of doing things which would reduce skylark numbers. Taking the extreme example of the World Trade Organisation agreements, free trade assumes an almost religious quality despite the fact that it is often socially unfair and environmentally unsustainable.

State intervention can go spectacularly wrong if the State, particularly when cooperating internationally, does the wrong thing. The EU policy on biofuels is a good example of a wrong-headed but very powerful State intervention. Supporting biofuel use across the EU leads to rainforest destruction and environmental degradation, but this initiative is so shared and now so embedded across all EU states that it is immensely difficult to dismantle. When the State gets it wrong it really matters.

If sustainable development takes account of the economic, social and environmental issues then under the Right, the economy gets the whip hand at the expense of social and environmental aspects, whereas under the Left it's the social aspects that may benefit at the expense of the economy and environment. It seems that it's either Profit before Environment or Jobs before Environment. Let me have men and women about me who are Green – and will give the environment full attention!

I sometimes think that I'd rather go for a walk in the country with someone from the Right. They would be more likely to appreciate it, but I would have to keep off the subject of politics. On the contrary, I'd rather talk politics with someone from the Left and put up with their having little feeling for the natural environment. That's a huge over simplification – but there is more than a little truth to it.

When we look at the major drivers of biodiversity loss and over exploitation then the role of markets, individuals and big companies is very clear. The problems of depleted fisheries, climate change (there are more climate change deniers from the Right than the Left) and habitat destruction (think of lost rainforests) all owe much to a free market approach. And it seems to me that the solutions lie in the tools of the Left, such as State intervention and international cooperation.

It seems to me that, more than in many other areas of public policy, the environment needs a benign government to take an interest. As I have written in this book – not much wildlife loss is caused by people deliberately intent on killing off wildlife. Most is the result of economic activities carried out in ways that harm wildlife almost by accident. I believe that the environment needs to be protected against wildlife loss, pollution and climate change, whether for its own benefit alone, or for our informed self-interest, and that these impacts are the result of unsustainable economic activity. I am a fan of regulation – yes, rules! Somehow the idea of simply telling people that they can't do some things, because they are bad things, seems to be going out of fashion with the Left as well as the Right. It's a good job that we've already decided to stop stuffing kids up chimneys to clean them because in these days we would be looking for economic incentives or a Big Society solution to this issue rather than banning it outright. That's my rant over – thank you, I feel better for that.

Right now

I am writing these words in January 2012 and they may not be worth reading by the time they appear in print in a real book in August this year. The current state of politics in the UK sees the nature conservation NGOs with very little political influence. Defra appears as a weak department with very little clout. Big decisions have been made by the Treasury (disproportionate cuts to Defra, and a review of the habitats regulations) and the Department of Communities and Local Government (weakening of the protection given to the natural world by the planning system) which will harm the natural world. The Chancellor, George Osborne, seems to be demonising environmental thinking as a brake on the UK economy and has ordered a review of the implementation of the habitats regulations. The Liberal members of the coalition government appear to have little environmental influence apart from the Secretary of State for Energy and Climate Change, Chris Huhne, who is the most unpopular member of the cabinet with Tory voters.

I cannot remember a time when those in political power appeared not just disinterested in the environment (which has often happened) but so actively anti-environmental. By the time that you read this book maybe this analysis will be proved wrong and the NGOs will have persuaded the government to adopt a greener stance, or the Liberals will have woken from slumber, or those green Tories will have re-established a grip on their party and the Government.

Conclusion

A lot of people say that they aren't interested in politics but they must be wrong. The thing is, if you leave politics alone then you are voting for them to do it to you, rather than you doing it to them. Life is politics – and wildlife is too.

Politicians have quite a lot of power, but nowhere near as much as they hope they will have when they walk into their first job. That's because the world is a complicated place and governed by rules of process and law which mean that you can't just walk in and start, for example, a badger cull simply because you are attracted by the idea. I think many Ministers find the life a bit frustrating. And it's not the role of NGOs to make their lives easy. There are plenty of other organisations out there being rocks for their interests and if you aren't being a hard place then the Minister isn't caught in a dilemma – and he or she will naturally take the easy way out.

All politicians prefer to be popular – it almost goes with the job since they are elected – and so the most powerful weapon we have is to make nature conservation a political issue. If only we could make more people take their love for nature into influencing how they cast their vote, then nature would get a better deal. I think we would see a much greater urgency amongst politicians of all parties to solve environmental problems if only those politicians thought that their own parliamentary seat, or that of a colleague, was at risk if they made the wrong decisions. I think it unlikely that environmental concerns will ever rise so far up the political agenda as to be crucial at election time, but I suggest that the NGOs get together and target those marginal seats where environmental issues might have greatest impact – perhaps those where the collective memberships of the NGOs is highest. It's time to target 30 or so seats in which to make the environment an issue – it might just capture the attention of all the political parties.

> *You have enemies? Good. That means you've stood up for something, sometime in your life.*
> **Winston Churchill**

CHAPTER 13

Advocacy in practice

A robin redbreast in a cage
Puts all heaven in a rage.
William Blake

This chapter discusses a number of mostly unrelated subjects which, while being interesting in themselves, also reveal something about how the world works and how influence can be exerted or thwarted. These are the insider stories that I can tell because I was there at the time and saw what happened – and in every case I wasn't just watching, I was involved in the day-to-day decisions, the policy development, signing off the press releases, talking to the media and planning the campaigns.

The trade in wild birds

The RSPB had been involved in campaigning against the trade in wild birds for many decades. At different times the issue had grown and subsided again in response to the events and players of the day, but in 2007 the EU banned the trade in wild birds because of fears regarding human and animal health. This move had been signalled in December 2006, rather to our surprise, by a letter to the RSPB from the then Prime Minister, Tony Blair, praising our campaign on this subject and promising to put the UK's weight behind a ban. Prime Minister Blair wrote that: 'the RSPB's campaign has graphically demonstrated that the catching and transportation of wild birds…causes unacceptable levels of suffering to the birds and can have a damaging impact on their wild populations.' It was nice of him to recognise that campaign but this was a noticeably greater engagement in nature conservation than we were accustomed to from Mr Blair – what was going on?

The arguments about animal welfare and nature conservation were almost irrelevant in stopping the trade in wild birds – it was the fear of bird flu that was behind this move. When an imported South American parrot died, apparently from bird flu, and fell off its perch in an Essex aviary, we recognised that there was the chance to get the bird trade banned for reasons completely different from those that had motivated the RSPB for so long. The fact that it was later revealed that the parrot probably did not have bird flu but that a batch of finches from Taiwan in an adjacent aviary were infected merely set the standard for confusion and error that came to characterise the bird flu issue.

The H5N1 virus originated in southeast Asia. It was known to be present in domestic poultry there and was commonly lethal in both domestic poultry and in people living in close contact with them. By the end of 2011 there had been 333 confirmed cases of human death from this virus worldwide although this probably underestimates the true number quite considerably. But to put this in context, and I hope this doesn't appear callous, there are about one and a quarter million car deaths across the world every year, so 10 times as many people are killed each day by cars than have died from bird flu in the last decade.

But that is in retrospect. Back in the early 2000s we were all worried that this new virus could sweep across the world and kill millions of us. And the disease did spread gradually westwards towards Europe. It was as though the Mongol hordes were inexorably sweeping across the plains of Asia to batter at the gates of your school, home or workplace.

That parrot, although it probably wasn't the parrot because it was really the Taiwanese finches, signalled the arrival of the disease in the UK and the fact that you could pop a few birds on a plane and have them breathing virus into our British air a few hours later was quite an eye-opener for the UK government. Was the trade in wild birds, with its conservation downsides and animal welfare issues, really such a good idea if it might also bring a killer disease to our shores? People started thinking that the answer was 'probably not' and the scale of the trade in birds rose up the political agenda while the trivial benefit to the UK economy of buying and selling parrots and finches became apparent.

Bird flu struck Europe with outbreaks at poultry farms but, thankfully, no human deaths. Wild birds were strongly implicated in transmitting the disease and suddenly people who knew something about bird migration were the flavour of the month. Suddenly the migration of the black-headed gull, wigeon and starling were of great interest to journalists, civil servants and politicians, who also wanted to know where these birds spent their time in the UK and how close they got to people on their stays.

Monitoring programmes were set up to spot suspicious deaths in wild bird populations and the public were asked to send in dead birds for analysis. In the RSPB, we were asked to join this monitoring effort and readily agreed to ask our wardens on a large number of nature reserves to devote time to collecting such data. Some other organisations were asked to do likewise, but refused. This monitoring was offered freely – it felt like something that we should do for the public good. But as time went by we asked for, and eventually got, a contract from government, although this was stopped years later as economic conditions tightened.

The H5N1 virus was first detected in a British wild bird in a long-dead swan in Cellardyke harbour in Fife. It says something about scientists' knowledge of viruses and birds that the microscopic virus was identified long before the swan; even though the swan's corpse was held in the lab (I hope they were wearing gloves!). It was at first assumed that the swan was a mute swan but later confirmed that it was a whooper – the implications for the arrival of the virus and its risk of onwards transmission being very different for the people-loving, bread-taking, primarily-resident mute swan compared with the migratory and 'probably heading to Iceland at the time' whooper.

But it was about a year later that the bird flu issue burst into an even greater media storm. I had spent the evening of Thursday 1 February in London having dinner with the then government Chief Scientist, Professor Sir David King, in the Orso restaurant near Covent Garden. We had talked about a whole range of issues as we enjoyed a good meal, including whether Sir David should recommend to Tony Blair that the UK should stock up on anti bird flu medicines.

The next evening as I pulled up outside my home, I heard on the 7 p.m. BBC Radio 4 news that there was a suspected outbreak of bird flu in a Bernard Matthews' turkey farm in Suffolk. Several colleagues had heard the same announcement and the evening was spent exchanging information and thinking about how the RSPB should react to this news and the inevitable finger of blame pointed at wild birds as having brought this disease to the UK. The eastern England location was certainly to be expected if wild birds were implicated, although an autumn arrival would perhaps be more likely than a mid-winter one. It was strange that although there had been an increase in cases of bird flu across Europe, most of them were still in eastern Europe and this outbreak represented a very big leap in the range of the virus. How had it made such a big leap? As is often the case, it was difficult to know – and we certainly weren't going to claim that wild birds were innocent since, on the face of it, they certainly might not be. On the next day, the Saturday, it was confirmed that the virus was the H5N1 strain and that thousands of turkeys were being slaughtered in an attempt to control the disease.

The RSPB joined a meeting of bird experts gathered together by Defra on the Monday morning and a good deal of interesting information began to emerge. Most interesting was the fact that the farm was also a turkey processing plant and that every week it received tons of partially processed turkey meat from Hungary. Hungary had suffered confirmed outbreaks of bird flu in its poultry farms and suddenly it seemed quite possible that migratory birds were not to blame for the leapfrog of the disease. Rather than worrying about how a diseased black-headed gull could have travelled across Europe without infecting other countries on the way, to end up at a turkey farm in Suffolk, it now seemed more plausible to imagine a lorry carrying infected turkey meat across Europe, from a country known to have suffered recent bird flu outbreaks, to the very farm where bird flu had been found in the UK.

David Miliband, the Secretary of State for Environment, Food and Rural Affairs at the time, had the job of making a statement to the House of Commons on that Monday afternoon where, given what we knew from his Department, we were surprised to hear him answer a question from his Conservative Shadow, Peter Ainsworth, by saying:

> *One of the hon. Gentleman's main points was about the causes of the outbreak, and he is absolutely right to say that getting to the root of it is a high priority. That is one reason why we have not dismissed any suggestions; we are pursuing all possible avenues of inquiry. It remains most likely that at the root of the problem there is a link with the wild bird population, but that does not mean that we should not pursue other avenues in a serious way, with the greatest of speed, and we are doing so. I will be happy to keep the hon. Gentleman and the House informed as the investigations by officials continue.*

But Ian Gibson, then the Labour MP for Norwich, followed up by asking:

> *My right hon. Friend makes the point that the most likely explanation for the cause of the outbreak is wild fowl. Is it not just as possible and just as likely that purchasing turkey chicks from*

> *Hungary might be an important factor? Has he investigated that to see whether it happens with British industries?*

He received the following reply:

> *There may have been one aspect of the question from Mr. Ainsworth that I did not answer, and it is linked – the so-called Hungarian connection. The chicks all came from within this country, so there is no Hungarian connection of that sort. The factory involved in the Hungarian outbreak was not a Bernard Matthews factory.*

I have often wondered how well-briefed Mr Miliband had been before he made that statement and whether he chose his words very carefully to focus on turkey chicks. If he had not used the words 'of that sort' then he would have been misleading the House of Commons although quite possibly not deliberately. And if Ian Gibson or Peter Ainsworth had mentioned turkey meat, although why would they, then would the answer have been very different? Was this a deliberate and knowing attempt to deflect blame onto wild birds or was it just one of those accidents which are very easy to make when a serious situation arises and everyone is worried – and most are confused?

We were surprised when we heard Mr Miliband's statement but expected the link to Hungary – a weekly lorry load of turkey meat – to emerge very quickly. In fact it was not until the *Observer* received leaked information from a Whitehall source, and published it on their website on the Thursday evening of 8 February, that the Government confirmed the Hungarian link and that they no longer believed the outbreak was caused by wild birds.

Bird flu was with us as an issue for several years and RSPB staff were often quoted in the newspapers, heard on the radio and seen on TV talking about it in what I think was a very sensible and rational way. We did get private praise from those in Defra who were trying to deal with the issue for the sensible and clear line that we took – never saying that wild birds were irrelevant to the spread of the disease, always recognising that the disease was of economic importance and a great worry to some farmers and never playing down the potential human health impacts. But we never allowed any hysterical nonsense about wild birds to go unchallenged either.

But you can see that, given the concern over the impacts of bird flu, it didn't take a genius to realise that shipping a lot of cage birds around, particularly when many of them were sourced in those very countries where bird flu was killing people, was a risk too far. That is why we got that Christmas message from Tony Blair in December 2006 and why the bird trade was banned at an EU level in 2007.

This is an interesting case study of advocacy. Going back to the mid-1970s the RSPB commissioned a study of the bird trade by Tim Inskipp. It was published in 1975, the same year he pointed out a roseate tern to me at the Patch at Dungeness, in a book titled *All heaven in a rage,* alluding to William Blake's poem. For many of us the case against the bird trade was a moral one – underpinned by nature conservation and animal welfare considerations. But, however convincing those arguments might have been to some, they had no political traction and ultimately it was human self-interest, economic and health impacts, and political calculations about the wisdom of dealing with the potentially dangerous bird trade, which finally led to that trade being banned. As is often the case, it's the economic and social aspects that are most important – not the environmental ones.

GM crops

There was a time when I seemed to spend more of my time thinking about genetically modified (GM) crops than about almost anything else. This was back in the late 1990s when firms such as Monsanto and Syngenta were looking for permission to grow genetically modified crops on a large scale in the British countryside. What was all the fuss about?

Well, the fuss, and there was an awful lot of fuss, was about the health and environmental safety of these crops and it stemmed from several different areas of thought.

The biotechnology companies like to remind us that all the crops we use are genetically modified – usually through the long, slow process of selective breeding where we choose particular characters to produce high-yielding or disease-resistant varieties of wheat or cattle or pets. The significant difference between that process, which relies on the natural genetic variation within a species, and the new GM crops, is that genes from completely different species, that never interbreed under normal circumstances, can be brought together by skilful scientists in the laboratory.

One example of this is the introduction of the scorpion venom gene into viruses so that the virus can attack pestilential caterpillars. However long you carry out selective breeding you will not get a virus to produce this venom gene but you may be able to introduce the gene artificially using complicated laboratory techniques. Other researchers have looked into the possibility of inserting genes from flounders into tomatoes, potatoes and tobacco to confer frost resistance, but nothing came of it. However, it is this type of potentially massive genetic change which makes the biotechnology companies interested in this type of technology and others worried about it.

Prince Charles got very worked up about GM crops but from the point of view of wildlife it is the impact of the GM crops that are of concern, not their means of production. I think that it is perfectly possible to imagine GM crops that are a boon to wildlife as well as ones that are terribly unfavourable. A crop that needed less inorganic fertiliser to grow, but was similar in all other respects, would be a leap forward for the sustainability of industrial agriculture in the UK. However, back in the late 1990s and early 2000s the GM crops proposed for commercial use in the UK countryside were herbicide-tolerant crops – ones which could tolerate (through genetic modification) the application of broad-spectrum herbicides to the crop.

It's probably worth explaining the limitations of current herbicides in UK agriculture in order to explain the attraction of GM herbicide-tolerant crops. If you are an arable farmer in the east of England you are probably making most of your money out of growing wheat. If you could, you'd like to grow wheat every year but what limits that is the build up of weeds in your wheat crop. Although you are using many very efficient herbicides, there are some weeds which have slipped through all your chemical defences. Most of the problem weeds are grasses, such as the species known to farmers as black grass but to botanists as slender foxtail.

Black grass is a common enough grass in the hedgerows and byways where I live in Northamptonshire – it's not rampant or dominant under normal situations but the herbicides we use at the moment are very good at killing off most other grasses in the crop and that allows both the wheat crop and the black grass to flourish. So far, no one has come up with a herbicide that kills off everything that the farmer would like to kill and so every few years it is wise to grow a break crop such as oil seed rape. Oil seed rape is not a grass – it's a broad-leaved plant – and so it is possible to use a different range of herbicides on fields where it is grown and this allows you to get rid of the black grass more effectively.

So, it would be great for the arable farmers of Cambridgeshire to have a magic bullet that killed every weed but left the crop standing. So a gene for resistance to herbicides, such as Monsanto's *Round-up*, would be potentially wonderful. It opens up the prospect of being able to grow whichever crop one wants, every year, in the same fields, without the worry of weeds reducing yields.

The GM crops under consideration were actually oil seed rape, sugar beet and maize, rather than wheat. The RSPB was worried that, because farmland wildlife depends on the small number of plants that slip through the herbicide net at the moment, the widespread use of such crops would be a dramatic change in UK agriculture – to the detriment of wildlife which already was having a really tough time.

The hero of this story, though there were villains too, was English Nature, led by their chair, the RSPB's ex-Chief Executive, Baroness Young of Old Scone. EN called for a delay in authorising GM crops until their ecological impacts had been tested. Right from the start the RSPB predicted that the farm-scale evaluations would prove worrying for farmland wildlife. To cut a long story short, we were right and the results of these trials were an important, though not the only, reason why GM crops were not approved for commercial release.

GM crops were unpopular for many reasons. Some feared that genetic modification could lead to super-crops that might become super-weeds themselves. Others feared that the herbicide-tolerant gene might transfer naturally to other species of plant which could then become new super-weeds. Organic growers feared 'contamination' of their crops with the GM varieties. The companies pushing these new crops were led by Monsanto, a company with an established unpopularity among those who remembered its Agent Orange herbicide, used by the US in the Vietnam war. And many just saw the technology as something that the US was trying to push onto a reluctant Europe.

It all got very heated and the pro-GM lobby was considered pushy, and to be exaggerating the environmental and agricultural efficacy of its products. On the other hand, the anti-GM lobby was described as a scare-monger and anti-science. Despite, I would say, being eminently reasonable on the subject, the RSPB was usually classed with the straight antis by some of the media, and certainly by the GM companies, because of our worries about the impacts of these crops on farmland wildlife.

Greenpeace, led by their Director Peter Melchett, were very anti-GM and actually took direct action by walking into a field trial of GM maize and, in front of the TV cameras, pulling up the plants by hand. They were soon arrested by the police and a while later I was subpoenaed to give evidence in their trial at Norwich Crown Court.

I had never set foot in a court room before and it was, I can clearly remember, one of the scariest things I have ever done – much scarier than giving a talk in front of 1,000 people, or doing a live interview with John Humphrys on the *Today* programme, or giving evidence before a Select Committee in Parliament. All those events are stressful and make the pulse run faster, as one wants to make a good impression (rather than look like an idiot) and do a good job for one's employer. But in the case of the trial there were 28 quite nice looking people standing in the Dock whose future depended on what happened in that courtroom in which I was a small player. That made it stressful like nothing else I have ever done.

My evidence concerned the value of the GM trial and took very little time. As I left the court a flurry of flashbulbs went off and journalists were shouting questions but they weren't for me. At the same time, in a nearby courtroom, the Fenland farmer Tony Martin's case was being heard for the murder of a burglar in his house. He was eventually acquitted as were the Greenpeace 28, but in their case only after a second trial, to which I had to return to give evidence again, after the first jury failed to arrive at a verdict. It still occasionally happens that someone comes up to me and starts by saying 'You won't remember me but I remember you…' and it is one of the Greenpeace 28 thanking me for my minute part in their acquittal, and I have become firm friends with Peter Melchett since that time, although if he gets at all uppity about anything I tell him that he might still be in prison were it not for me!

As well as the farm-scale trials to test the ecological impacts of the particular crops proposed for commercial license, the government also set up a Science Review Group under the Chairmanship of the Government Chief Scientist at the time, Professor Sir David King. This large group was to look at the general concerns about GM crops and I was asked to join the panel, mostly comprised of eminent scientists involved in GM science and apparently mostly in favour of GM technology. Sir David chaired the meetings very ably and they were held in public – although very few of the public bothered to attend any but the first meeting.

The Panel looked at three main issues: food safety issues; super-weed issues; and the broader ecological issues of GM technology. Meeting every few weeks for several months, the Panel produced two long reports on the subject which essentially said – it depends on the exact specification of the GM crop. Not perhaps the most exciting of answers but probably the right one.

I was an active member of the Panel and learned a lot about how to influence a report of this type, in this type of setting. Here are some tips on how to gain influence if you are ever in a similar situation:

- Attend each and every meeting for each and every minute of the meeting. If you are absent then changes can be made and you can't really complain if you had the opportunity to be there but left early for that train home.
- Don't talk too much but do speak out – that's what you are there for and you can't rely on others to know what you know or to express it as well as you.
- Listen carefully to what others say and try to refer to their points when making your own.
- If the meeting becomes adversarial, then be prepared to acknowledge the truth of the other side's points sometimes. If you are a reasonable person then this helps to prove it – and if you aren't then it might fool people into thinking that you are.
- Be nice to the secretariat staff. There are often several civil servants involved, who run the show from behind the scenes, some senior but some of them quite junior. They organise the rooms, lunches and coffees, and may be responsible for paying your travel expenses if they are being reimbursed. They have more access than you to the Chair of the group outside the meetings and when it comes to the final report they may well have a hand in editing and producing it. It's just common courtesy to be polite to these people but it might also work to your advantage now and again.
- Volunteer for jobs. There are always jobs handed out in such circumstances and you should show your commitment to the group by volunteering to do some of them.
- Work assiduously on the report. Choose which areas you want to influence and always comment on the exact wording proposed – and if you don't like it explain why and suggest an alternative form of words.
- It's not over until it's over – anything can be changed at the last moment

and things often are. Remain vigilant and if you have had to make a point 20 times before to get the wording right, if you don't make that point the 21st time it is your fault that the wording is wrong – you have been 'out-endurance'd' in the run to final publication.

It would be tempting to claim that the herbicide tolerant GM crop battle was won through science and rational argument but it is never that straightforward. Politicians often fall back on the science as their reason for action or inaction while their real motives may be much more complicated. GM technology was unpopular and the industry in favour of it handled almost every aspect of public relations badly. Every impression was of the nasty industry bullies against the plucky organic farmers and fearless environmental campaigners, such as Friends of the Earth, Greenpeace and the RSPB. Many of the public distrusted the industry spokespersons and the technology seemed like such an American import that many also suspected Tony Blair of just wanting to humour his mate Bill Clinton.

However, even if it wasn't just straight science that influenced the decision to deny commercial approval for these crops it certainly wasn't the triumph of scare-mongering over science that the industry has tended to claim in retrospect. There were some dodgy and outrageous claims about GM crops and their safety and value to mankind made at the time.

My personal belief is that GM crops will probably play a future role in delivering a more sustainable agriculture, which produces more safe food with a lower environmental impact. But as yet, I haven't seen the crop that is going to do that and, personally, I am a bit sceptical about whether the research arms of massive global pesticides companies are the best places to develop those useful GM crops. This strikes me as the sort of area where governments should get together and fund research to develop such crops and if successful roll them out appropriately. But that's such a utopian hope I doubt it could ever happen. So we wait for the likes of Monsanto to bring forth a GM crop that *will* save the world – it might happen, but I am not holding my breath.

Lead ammunition

Lead is a poison for just about all life on Earth – it's not good for you or other mammals or birds. On human health grounds we have removed lead from paint, water pipes and petrol during my life time, reducing our risk of illnesses to organs such as the heart, kidneys and intestines.

When I was based at Oxford I sometimes helped researchers studying the impacts of lead poisoning on mute swans. This research was led by Prof. Chris Perrins but involved a range of students and post-docs including Jane Sears (who later moved to the RSPB), Phil Bacon and Mike Birkhead (who moved on to making nature films).

The study, which was part-funded by the RSPB, involved looking at how, and how many, lead fishing weights were eaten by swans, and what impacts that had. At the time my involvement was catching swans and holding them still while someone else took blood samples from their legs.

Now everyone 'knows' that a swan can break your arm, leg or any other limb with one waft of its powerful wing (an urban or rural myth) so this must have been a very dangerous task – not at all. Catching swans on the Thames and its tributaries involved driving around attractive places, stopping off at country pubs, mucking about on the river with canoes, getting wet on hot summer days and a bit of energetic swan-wrestling. I learned to use a swan hook to catch a bird around its neck and pull it, dangerous wings flapping, quickly towards me, then to gather up the swan in my arms.

Swans live in places where lots of people also live, so some of our swan wrestling was carried out in public places. One soon got used to this and it was an eye-opener onto British reserve. I remember arriving one sunny day at a full car park by the Thames while the occupants and a few families fed the ducks. It was a quiet, peaceful scene. Three of us got out of our white van, walked up to a family of dirty brown cygnets with their two parents and each grabbed two of the young swans by the neck, picked them up and walked off back to our van whilst the parents hissed and spread their wings and attacked us. The peaceful scene was transformed into mayhem. We tied the cygnets' legs and wings up so that we could line them up in the car park without them wandering off and then proceeded to weigh and measure them, ring them and take blood samples. The assembled families and resting commercial travellers watched in amazement but none asked us what we were doing or whether we were allowed to behave in this way. Eyes peered from behind the twitching pages of the *Daily Telegraph* in a nearby BMW but its occupant watched without a word.

Those studies, with which I was only marginally but enjoyably involved, showed that many swans ingested discarded lead fishing weights while sifting the gravel on the river bed. Swans with lead poisoning often died a painful and unpleasant death as the poison paralysed their oesophaguses so that they could not swallow their food. They kept eating through hunger but less and less food was properly ingested. Their necks became compacted with food and you could tell a swan in the later stages of lead poisoning by their sad behaviour and also by their bulging necks.

Just one large lead weight was enough to poison a swan but most acquired their lead loads from picking up many small lead weights from the river sediment over a long period of time. There was no doubt that lead was bad for swans, as it is bad for us all, and that accidentally or deliberately discarded fishing weights were the source of this poison.

Lead is a useful metal as it can be easily bent by hand and that's why it has been used for fishing weights. I remember buying split lead weights and biting them shut on my line as I fished for minnows and gudgeon – and swallowed a few of them too, I don't doubt. Lead fishing weights were, almost entirely, banned from use as a result of this research but not until after fishermen had argued that there was no impact of lead on swans and that even if there was then lead weights were so integral to their pastime that they could not sit by the riverbank without the help of this toxic heavy metal. This was, of course, utter rubbish but illustrates the type of argument that is always produced when any type of social or environmental change is proposed.

Similar views were expressed by motor manufacturers when it was decided to remove lead from petrol. Its anti-knock properties, we were told, could not easily be replaced and there was a danger that road transport would become much more expensive and less feasible if lead were not used. Nonsense again!

And when we were involved in getting lead gunshot banned when shooting over wetlands, similar arguments were made by the shooting community. They claimed that there were no viable alternatives to lead ammunition (there are), that lead poisoning is not a problem for wildlife (it is), and that a change to lead-free ammunition would require changes to guns and to shooting techniques (sometimes true).

In the late 1990s and early 2000s the RSPB worked with others to secure a ban of the use of lead ammunition. It had been established that lead poisoning was a major source of mortality for waterfowl, which ingest the lead shotgun pellets that accumulate in wetland sediments – as well as for some birds of prey, which ingest the lead from the carcasses of their waterfowl prey. This ban applied only to the shooting of wetland species over wetlands, the hunting lobby having persuaded the then minister, Michael Meacher, that there would be considerable opposition to a total ban (maybe from all those grouse-shooting peers in the House of Lords) and that he should be content with a ban over wetlands. I remember Mr Meacher 'phoning me to ask whether the RSPB would accept this compromise or whether we would oppose it – we decided to accept it as progress and bank it as a step in the right direction.

Since then it has been illegal to use lead ammunition to shoot ducks in England (of course, the situation is slightly different elsewhere in the UK) and yet we know that non-compliance is still high. Prince Harry and his two companions were questioned about their use of lead ammunition when duck-shooting at Sandringham in 2007 and the Crown Prosecution Service were quoted in the *Guardian* as saying:

> *The question of whether cartridges containing lead shot were used in breach of environmental protection regulations was considered, but as the bodies of wildfowl which were shot had been removed from the scene by the time the police arrived, it is not possible to say if this was the case. The three suspects denied any breach of the regulations.'*

But a study of ducks bought in game-dealers, supermarkets and butchers carried out by the Wildfowl and Wetlands Trust and the British Association for Shooting and Conservation in 2010 has shown ample non-compliance with this law. Seventy percent of the ducks had been shot illegally with lead – much the same proportion as a study carried out eight years earlier. It's not as though this illegality is a result of ignorance – interviews showed that shooters mostly knew what the law was but almost half of them (45%) admitted to breaking it. Their reasons ranged from not believing that they would be caught, not believing that lead was a problem or simply not wanting to pay a bit extra for non-toxic ammunition – that is, criminality, ignorance and meanness.

Meanwhile, the evidence has grown that lead is not only an important and avoidable problem for wildlife, but a bit more of a health hazard than we had previously appreciated.

The European Food Safety Agency has concluded that certain groups of people are at some health risk from lead ingestion – those who eat a lot of game and perhaps young children and foetuses. Health experts now essentially say that there is no level of exposure to lead that is safe – every bit of lead that you ingest does you some harm – so it would be sensible to reduce those levels wherever possible.

One of the most interesting and surprising facts on this subject came to my attention when a group of conservationists from the Idaho-based Peregrine Fund visited the UK and talked to us about the reasons for vulture decline in India, Pakistan and Nepal. They had discovered that the veterinary use of a non-steroidal, anti-inflammatory drug, diclophenac, was at the root of the vulture decline. But after talking over that important piece of work the then Chair of the Peregrine Fund, the late Bill Burnham, showed us an X-ray of a deer carcase that had been shot with a lead bullet.

The bullet track was obvious on the X-ray, but Bill also pointed out a large number of bright white spots and explained that these were tiny lead fragments that had come off the bullet and distributed themselves throughout the body of the deer. More such evidence was presented at a wildlife conference in 2008 and we at the RSPB made similar examinations of deer which had been shot at Abernethy, with the same results. Lead ammunition, in the form of bullets and shot, sheds tiny lead shards throughout the carcase – you can't see them and it's just about impossible to remove them through butchering. And of course for a scavenger such as a fox, crow or golden eagle these tiny bits of lead are ingested.

This information persuaded the RSPB to switch to copper bullets for any deer-culling on its nature reserves and highlighted the need for generally more stringent controls of lead ammunition.

Further evidence came from a WWT study which showed that normal cooking of game bought in shops often produced lead levels in the meat that were way above those that would be allowed for beef, pork, chicken or lamb. This is all very inconvenient for the game shooting industry, which has been campaigning for many years to persuade more people to eat game whilst being completely aware of these studies which highlight the animal and human health impacts of lead ammunition.

It's surely time to stop making a fuss about it and simply completely ban the use of lead ammunition, just as many other countries, and some USA states, have already done. If Denmark can cope without lead ammunition – then surely so can we.

I still eat some game that has been shot with lead – it's my choice and I am quite well-informed about the risk. I do wonder though what the impact of a much higher game intake might be on a gamekeeper's children whose father has a ready supply of cheap, lead-shot meat.

Two of the symptoms of lead poisoning are apparently confusion and irritability. I've often remembered this whenever a red-faced man has spluttered at me that he has, '…eaten game shot with lead for years and it's never done me any harm!' Really? – are you sure?

Biofuels

When we burn coal we are tapping into the energy of creatures that lived on Earth 300 million years ago. That energy, derived from the sunshine all that time ago, has been trapped underground in coal deposits and is released in our power stations or home fires all these years later. By burning coal, oil and gas we are suddenly putting back into the atmosphere large amounts of carbon dioxide that had been removed from it millions of years ago.

So why not stop doing that and instead use the energy trapped in today's plants for energy and replant those plants we have used (far easier than trying to recreate coal as quickly as we burn it). That would mean we are simply recycling the carbon all the time and not adding extra to the atmosphere – and therefore not exacerbating climate change.

The idea sounds like a good one and in many ways it is. Using a renewable resource is certainly a step forward compared with using fossil fuels in terms of carbon production. And the term biofuel makes these fuels sound natural and green and good for the planet. But life is more complicated than that – I'm sorry but it is.

Governments across the world have been won over to the idea of biofuels and here in Europe it is now an EU-wide policy that when you fill up your car with petrol or diesel it contains a fixed percentage of biofuel, most probably derived from palm oil but possibly from maize, soya or some other plant.

Depending on the means of production, biofuels are estimated to save between none and around 80% of the carbon that would be produced by using fossil fuels instead. That makes it sound as though all we have to do is choose the right biofuels and we have made a big leap forward. If only life were that simple.

There are difficulties in assessing the true greenhouse gas emissions involved with growing and processing biofuels, and many, including the Nobel Prize-winning Professor Paul Crutzen, believe that the energy savings of biofuel use have been overestimated because they do not take into account the impacts of the nitrous oxide produced as a result of using fertilisers in crop production.

However, to my mind, the biggest problem with biofuels is that the world is a small, crowded place. It takes a lot of land to grow crops – after all 75% or so of the UK's land area is farmland devoted to livestock or crops. If we could reduce the area of land under agriculture, then presumably we would – we'd have much more natural habitat to go out and enjoy and organisations like the RSPB would have rather little to do – hooray! Because the world is a small, crowded place if we need land to grow biofuels as well as food, then we need some more land. Unless we are going to stop eating, and this is a small, crowded planet with a lot of hungry people on it, we'll need to use land not currently used for food production on which to grow biofuels.

The consequences of wanting more land for commercial use are the same now as they have always been – we end up cutting down rainforests, ploughing up grasslands and draining wetlands. This process, which has been going on ever since mankind invented farming, has previously been driven by the economic value of food production. Now it is driven by the economic value of fuel production. This is terribly bad news for tigers, orang-utans and a whole host of other less charismatic creatures, but it is also madness in carbon terms.

The world's rainforests are massive carbon stores. Something like 14% of climate change is caused by deforestation. And so to calculate the true energy balance of biofuels, one has to take into account the destruction of rainforests and other carbon-rich habitats to provide land for their production. When you do these sums (they aren't very difficult but I will spare you the details) you find that even a tiny amount of rainforest destruction to enable biofuel production will negate the energy savings from these so-called green fuels.

I made these points in spring 2008, on the RSPB's behalf, to an enquiry into biofuels which was ordered by the Secretary of State for Transport, Ruth Kelly, and carried out by Professor Ed Gallagher, formerly a senior staff member of the Environment Agency.

The Gallagher report clearly signalled the grave danger that biofuels would provide little or no savings in terms of greenhouse gas emissions, but would cause rainforest destruction, biodiversity loss and a host of social problems as well. But the report recommended a slowing down of the growing use of biofuels, rather than a ban on their use – as the RSPB and many others had called for. At the time I was quoted as saying that most people know what they should do when they are in a hole, but the Gallagher report recommends carrying on digging – just dig more slowly.

The RSPB played a large role in the general debate over biofuels and I think we could claim a leading role in the energy and biodiversity arguments. We had lobbied government about our concerns and also taken them to the EU, where the Biofuels Directive finally imposed an EU-wide obligation to use biofuel as an additive to oil-based transport fuels. We produced a newspaper advert that showed threatened wildlife and asked the public to email Ruth Kelly with their concerns, which resulted in her inbox crashing.

Let us revisit the conflict between using land for food or fuel production. Already in the USA a quarter of the maize grown now goes into biofuel instead of foodstuffs, compared to a decade ago when biofuels were practically non-existent. This pushes up the price of food across the world which hits the world's poor hardest. The fuel v food debate is a critical one. It's worth remembering that in the UK the National Farmers Union strongly supports biofuel production and also maintains that food production should override biodiversity protection in public policy. In the real world it is difficult to support food production *and* biofuel production at the same time without looking hypocritical or stupid.

All the problems associated with a shift to biofuels are understood by governments across the world and yet far too little is being done to address them. In the UK we have something called the Renewable Transport Fuels Obligation which requires that by 2013 all the fuel you put into your car must be 5% biofuel. This requirement stems from an EU-wide Biofuels Directive so something similar applies to every German, Italian and Pole filling up their cars, lorries or tractors. Every time I fill up at a petrol station I fume at the fact that I may be pouring bits of tiger into my tank by way of destroyed rainforest – and that I can't go to another petrol station for non-biofuel fuel, even if I were prepared to pay more. I am trapped in a flawed policy and have no choice.

But the decision-makers do have a choice – and they are not taking it. Although you could argue with the details of the case as summarised here, it is interesting that practically everyone essentially agrees with it. I have been told by civil servants in London and Brussels that this argument is right, after which they shrug and say 'but we can't do much about it'. Policy is sometimes made in a hurry and then takes a very long time to reverse. There are now signals and agreements sent to industry here and overseas which apparently are more important than doing the right thing. It is ridiculous, and many look embarrassed and shifty about it, but little progress is made in fixing it. Is there too much loss of face in coming to our senses?

I would like to see the biofuels obligation removed completely. The influence of the RSPB and others on the Gallagher Review led to the UK slowing down its adoption of biofuels and the review clearly stated that further adoption of biofuels should be accompanied by policy safeguards to ensure sustainable production and reduce loss of natural and agricultural habitats. But, how do you put policies in place which protect you from unintended consequences? One of the lessons of the biofuels debate is that the effectiveness of public policy has its limits.

When the EU sets a biofuel target for transport fuels it creates an economic incentive across the world for biofuel production. So a farmer in Iowa might sell his corn for ethanol production rather than food. He doesn't cut down a rainforest but by removing his corn from the food chain he creates a need for food production that must be met somewhere else in the world. In Indonesia a small patch of rainforest may be illegally logged for its timber but also for the food production value of the land. The cleared rainforest doesn't necessarily produce biofuel (although it might) but it is a knock-on effect of a global biofuel need. It is beyond our wit as a species, with the policy instruments at our command, to send out a signal of 'more biofuel' and to ensure that that does not mean 'less rainforest', 'less food' or 'fewer tigers'? The tragedy is that if it does mean 'less rainforest' then the main policy reason for saying 'more biofuel' in the first place – reduction in greenhouse gas emissions – is seriously undercut.

Ed Gallagher once said to me, after his report, that the evidence that I gave for the RSPB had been very compelling and had influenced the report greatly. This was very good to hear because I had spent the whole weekend before the Monday when I gave evidence trying to think of how to get the arguments across in a truthful, clear and convincing way. So maybe this was a small victory for the RSPB but also for common sense. Maybe we made a difference. It's sometimes difficult to know and it's sometimes difficult to feel that it's a big enough difference, but if we don't try then we are failing in our duty. We can't guarantee victory in environmental battles but it is important to be on the right side of the debate.

Peat

As a teenager I would sometimes visit the Somerset Levels on a Sunday afternoon drive with my parents. In winter I'd be on the look out for flooded fields with snipe, golden plover, lapwings and the occasional dunlin or ruff. Sometimes there would be big floods with Bewick's swans and many ducks, and there was always the chance of a merlin, hen harrier or peregrine.

The flat landscape of the Somerset Levels sits on peat deposits laid down after the last Ice Age and while I was looking for birds my parents would sometimes buy a bag of peat for the garden – what could be more natural than using the Somerset soil to make our garden more beautiful?

We would often buy peat very close to what is now the RSPB nature reserve at Ham Wall, a wetland created on the site of old peat workings. The reserve and the surrounding Levels now hold a variety of wetland species that I never dreamed of in my youth – bitterns, little egrets and grey herons, but also little bitterns, cattle egrets, great white egrets and always the chance of purple herons in the future. Our peat purchases helped to create the hole that is now filling up with herons.

This story of eventual nature conservation gain from peat extraction is an exception to the usual rule. Peat removal for horticultural use usually leads to the destruction of lowland peat bogs, often of priceless wildlife value. Thorne and Hatfield moors in north Lincolnshire and south Yorkshire, and the Solway peat mosses, such as Bowness Common, contain most of the English lowland raised peat bogs and all these remaining, but damaged, sites now have some form of statutory protection. Peat extraction depends on a company having the mineral rights to a site and in the past these have been granted on long leases. It is understandable that if your business acquired the rights to extract peat in the 1950s or 1960s, and built its commercial enterprise on the assumption that peat could be extracted until the licence expires in the 2010s or 2020s, then you will not welcome anyone who comes along midway through this period and tries to cramp your style.

Peat extraction not only destroys the complex ecosystem of a natural peat bog but also, as we now know, generates greenhouse gases. Peat is similar to coal in some ways – formed from the bodies of dead plants it is rich in carbon and represents the stored energy of the sun from thousands of years ago. Although both peat and coal are being created as we speak, the process is so slow that we can essentially think of them as being non-renewable resources.

Ireland's peat-fired power stations and the peat fires of Shetland's crofters are throwing carbon up into the atmosphere in a similar way to their coal equivalents. But surely our bag of peat for the garden had no such impact?

I'm afraid that it did. If you let sleeping bogs lie then their carbon remains tied up in the soil but once you start digging it up and drying it out then it starts to break down and release all that stored carbon into the atmosphere. So a bag of peat not only impoverishes the peatland habitat from which it was collected (in this country or abroad), but also increases carbon emissions in the garden to which it was taken – please think about that the next time you consider buying peat.

There are alternatives to peat, and gardeners and horticulturalists are a bit divided about their efficacy. Of course, just because they are divided about whether non-peat products work properly, it doesn't mean that the problems of peat extraction are irrelevant. I'm no gardener – I am more likely to damage than nurture a garden – but I think that enough people say there are very few cases where the use of peat is essential for me to believe them.

Several nature conservation organisations have campaigned against peat use, on and off, for years and with some success. We know that many RSPB members have stopped using peat but we also know that many have not. Government has long favoured a protracted voluntary approach which relies on companies gradually changing their ways and people being so well informed that they are sufficiently motivated to change their ways. An approach that is unlikely to succeed anytime soon.

And government ministers always listen to the cries of industries that extract or use peat that jobs are at risk if peat use is banned or restricted, but don't heed the quieter voices of the companies which produce peat-free products whose businesses would take up the slack if it were.

Many companies selling peat have diversified into peat-free alternatives as a business move. This enables them to look greener but probably more importantly it gives them a foot in both camps – whether peat stays or goes they are positioned to exploit the market. This makes good business sense. But as in so many areas you can't expect business to be lobbying for change when that change may not suit them – business will often position itself to react to change that might be in the wind, but rarely calls for it. Or if it does, it is the new progressive companies that call for change and their voices are lost in the larger, more established voices that may well be part of the problem in the first place. A common situation and the reason that governments need to take the lead. But in the case of peat, a small issue in the bigger scheme of things, governments have not yet had the courage to act.

Back to forestry

Forestry has played a surprisingly large part in my professional life. As I told you in Chapter 1, I got into the RSPB through finding a mistake in Colin Bibby's work whilst writing a small book for the Forestry Commission on birds and forestry which was eventually published as *FC Occasional Paper 26* with Steve Petty.

After working in the Flow Country I was steeped in the big forestry issue of the day – upland afforestation – and I was approached by Andy Richford, commissioning editor for T&AD Poyser, to write a book on forestry and birds with my friend Roderick Leslie. Roderick was a forester working for the FC, but also a birder and ringer, and, actually, an RSPB Council member too, so I guess Andy thought that between the two of us we had a fair grasp of the subject.

Birds and Forestry was published in 1989 and dealt with the forestry issues of the day. It concentrated far more on whether and where to plant trees than any other single issue. Our argument was that forestry wasn't all good but it certainly wasn't all bad and that the balance of conservation benefit depended on how much afforestation took place and where it happened. Forestry in the right places could actually do some good but smothering places like the Flow Country with industrial tree-farms was a terrible idea.

There was a certain amount of nervousness in the RSPB about my writing an opinionated book on a subject as contentious as forestry, but I agreed to let the RSPB's forestry policy officer, Ian Bainbridge, read the draft in advance and that smoothed the way. When the book was published it didn't take long before the RSPB was claiming that it represented the Society's policy on forestry.

Most people are generally in favour of trees, which has always seemed a bit prejudiced to me. Being in favour of trees seems just as prejudiced as being against them – surely, it all depends? Ancient forests such as Abernethy, the New Forest and the Forest of Dean have all been shaped by humankind over the centuries, but in some respects still resemble natural ecosystems. Their ancient nature is reflected in their flora and invertebrate fauna far more strongly than in their bird populations. Such forests are both scenically attractive and biologically important.

At the other end of the scale, a plantation of non-native conifers, planted in rows and rapidly shading out the ground vegetation is a uniform crop as much as a field of winter wheat. If planted in the wrong place it may destroy a lot of existing biodiversity and put back very little.

So, forests are like farmers, footballers and fishermen – all are different, some are fabulous and some are foul.

Woodland birds are in decline, and if it weren't for the sorry state of farmland birds they would be the focus of much more conservation attention. The populations of species such as the willow tit, spotted flycatcher, wood warbler, lesser-spotted woodpecker, hawfinch and tree pipit have all declined by over 50% in the last few decades and all feature on the current Red List. These species don't seem to have much in common and, so far, we are struggling to blame their declines on any general change in the environment in the way that 'intensification' can explain farmland bird declines.

This may be because there is no single underlying theme – there needn't be, after all. However, it is a good way to start looking at such issues – take a broad perspective over time and space and think what might be going on. But before I sketch out two major themes that may explain things, let me point out an alternative way of looking at it.

Some people have leapt to the conclusion that because woodland birds are in decline and because the non-native American grey squirrel lives in woods – and eats birds and their eggs – then it must be the spread of these squirrels which is to blame. This isn't, at first glance, a daft idea as non-native predators often do cause problems for wildlife, and squirrels do not just eat nuts, they truly are predators. Added to which, the declines in songbird numbers do tend to be worse in the southeast part of the country, where squirrels are numerous.

But the broader perspective is useful here – many of the species of woodland birds that are declining in the UK are declining across Europe as a whole – and, apart from a small part of northern Italy, grey squirrels don't occur elsewhere in Europe. So however much you want to demonise the grey squirrel for causing the decline in our native red squirrel, you can't demonise it for causing spotted flycatcher declines –unless there are different causes at play elsewhere in the spotted flycatcher's European range.

But the spotted flycatcher is just one of many declining European woodland, and non-woodland, species that are trans-Saharan migrants. Perhaps there are some climate related, or insect-related (since most summer migrants are insectivorous) or Africa-related issues which are affecting many woodland species.

After the migration related themes, the second general issue that almost certainly affects woodland birds is the lack of woodland management. We sometimes forget that our woodlands have always been used for grazing, timber production, firewood, coppicing etc. and that those activities have helped to shape them and their wildlife.

Here again, as almost always, a wider perspective is useful. Woodland plants and butterflies have also declined and the suggestion is, often based on good studies, that lack of management lies at the heart of these declines too. So maybe the bluebell, the lesser-spotted woodpecker and the wood white butterfly are all suffering from the same changes in woodland management and that by restoring management to our woodlands we could spark a revival in woodland wildlife?

This is an attractive prospect and certainly has implications for how nature reserves and state-owned forests should be managed. But many of our woodlands are privately owned, sometimes by pheasant shooting interests, and how would we persuade these owners to carry out more management such as selective thinning of trees? The answer might be in the value of wood as a renewable fuel supply. This might really be a win-win-win situation in our existing woodlands. Harvesting biomass from woodland on an entirely sustainable basis may provide a useful form of renewable energy (which doesn't replace food production) and an economic incentive for management while incidentally recreating some of the conditions that benefit woodland wildlife. It's worth a try and I'd like to see some large-scale experimental demonstration sites to give it a fair test.

The most recent debate concerning forestry was stimulated by the coalition Government's proposals to sell off some state-owned forests to raise money and reduce the size of the state bureaucracy. Its proposals went further than many had expected and the public backlash, whipped up partly by the recently established lobby group 38 Degrees, was impressive. David Cameron cut his Secretary of State for Environment, Caroline Spelman, off at the knees when he was asked whether he was happy with his own government's proposals on forestry and said that he wasn't. Since then not much has happened that we can see, as politicians and civil servants lick their wounds and an expert group cogitates on the future of these forests.

It's tempting to be drawn into thinking that you have to be either for against the government's proposals, and it's sometimes tempting to decide without actually understanding what those proposals are (at least some commentators seem to be tempted in that way), but the broader view is worth taking. Why does the government own forestry plantations when it doesn't own farms or fishing fleets? The answer was that timber for pit props was considered a strategically important resource when the FC was set up in 1919 – but we need a better reason than that now.

To understand this issue I believe we have to return to the fact that all forests are different – and that some are fabulous and some are foul. Some are quite simply sites for timber production and while there is nothing wrong with that, most such sites could be managed perfectly well by private enterprise – who might pay us, the taxpayer, for the privilege of making money, which would be a useful addition to the nation's coffers at the moment. On the other hand, places like the New Forest and the Forest of Dean are not primarily economic timber farms. Their values are difficult to express in monetary terms – they have cultural significance and are beautiful places for people to walk and for wildlife to live.

Under these circumstances I can't see that disposal of forests, if they are the right forests (the fouler ones) is a bad thing at all. Why should the government be involved in a small fraction of UK timber production when they don't dig coal out of government-owned coal mines held up with government-grown pit-props? Personally, I believe that there is probably a better case to be made for government to own land to grow food in a sustainable way. I'd be happier with state farms than commercial state forests. While you always have to start from the *status quo* – you should really be asking 'is this where I would be if I had a free choice?' The answer to the commercial forestry question, for me, is no.

But there is a perfectly good case for the fabulous forests to be managed by the state. Their future is not to generate economic wealth, but cultural wealth and since that is something that it is hard to make money from, but is important in our lives, then there is a role for government there.

If, after mature consideration, government chose to continue with the disposal of some state-owned forests I wouldn't object provided that sensible safeguards regarding access and environmental standards were attached – and that they were sold at a decent price. If this happened in the way that I envisage, then the remaining forests in state ownership would become fewer in number and more focussed on high quality forests. Under these circumstances the role of the FC (which is, by the way, a non-Ministerial government department – something as rare as swallows in December) would come under question – and that would be a good thing.

A thinking conservationist would neither berate the FC for damaging the environment nor laud it as a peerless champion for wildlife. The truth lies in between, as it so often does in the real world. The fact of the matter is that the FC's remit is a mixture of commercial timber production and delivery of a public service to people in the countryside. If we sold off the most commercial bits of the timber business to people whose real business was growing trees then what was left of the FC could form a very well-informed and able part of a new statutory agency engaged in delivering wildlife, access and well managed countryside to us all. Merging the FC with Natural England could produce a body with real ability to deliver a much better future for woodland wildlife. Its practical expertise, focused on our very best forests, would deliver the thriving woodland wildlife seen in so many of our best managed, publicly owned quality forests.

Conclusion

This chapter has told some of the advocacy story regarding the bird trade, genetically modified crops, lead ammunition, biofuels, peat and forestry. In every case some progress has been made and in no case has victory been secured – that's life. But in no case is the war over – there is still scope for making the world a better place and winning the arguments. I was once asked what my exit strategy was from a particular course of action and, after a moment's thought, suggested that 'victory' would be a good option – otherwise known as 'driving the enemy into the sea'.

But final victory on environmental issues often takes a long time and is only won through stubbornness and determination – and as with military victories one wishes for lucky generals. There are almost always vested interests arrayed against you – otherwise you would have won already and nature would be having a much better time of it. The industry often talks up its case outrageously whether it be the economic benefits of GM crops or the economic downside of not being able to use your Purdey shotgun with steel shot.

That something as cruel and economically insignificant as the bird trade could continue for so long is amazing – and yet it was not the conservation or animal welfare arguments that won the day, it was politicians' fears that they would look foolish if they continued to allow birds to be shipped around the world when bird flu was spreading and might leap across the species barrier creating a human pandemic.

It's difficult to understand why peat is still sold in garden centres when its climate and habitat impacts are so clear and its horticultural necessity so low.

GM crops were quite wrongly hailed as a massive technological step forward and again their benefits to anyone – the public and farmers – were exaggerated and their problems glossed over by the industry that had most to gain from them financially.

Forestry has remained in a bit of a *cul de sac*, separate from other land use issues because of history rather than logic, and much remains to be done.

My advice to environmental campaigners would be to make sure of your own facts but also to look very hard at what the other side is saying. And don't believe that facts alone will win the day. A mixture of personal prejudice, belief, luck and politics will determine whether the outcomes are what you seek or what you dread.

Once, at the end of a rather testy meeting between a bunch of NGO climate people and the then Energy and Climate Change Secretary, Ed Miliband, I tried to end the meeting on a friendly note. I pointed out to Ed that although he'd seemed a bit irritated with us, and no doubt we had seemed a bit irritated with him, we were on the same side and that, unlike almost everyone else he spoke to on these issues, we weren't there with a vested interest. Greenpeace, Friends of the Earth, WWF and the RSPB wouldn't make money whichever route he travelled. We were there sticking up for what we believed was right and so he should take a bit more notice of us than he did of the industry representatives that would spin him a line on jobs and the economy. The sentiment put a smile on Mr Miliband's face and helped end the meeting on a good note – but it is also true.

> *All endeavor calls for the ability to tramp the last mile, shape the last plan, endure the last hours toil. The fight to the finish spirit is the one... characteristic we must possess if we are to face the future as finishers.*
>
> **Henry David Thoreau**

Chapter 14

Snippets

Be amusing: never tell unkind stories; above all, never tell long ones.
Benjamin Disraeli

This chapter is different – it's a selection of anecdotes that I hope that you find interesting but that don't fit in anywhere else. It also forms a bit of a break before the last three chapters which are more forward looking compared to their predecessors.

The anecdotes here are deliberately jumbled up and in no particular order.

Challenged by Paxo

The RSPB entered a team for *University Challenge: The Professionals* one year and I was on it. I got to say 'Nuts!' to Jeremy Paxman but we were out-gunned by The Imperial War Museum and in the end did badly. Jeremy Paxman talked to us briefly and said 'RSPB - I'm sure there's some reason why I'm not supposed to like you' and I suggested that it might be because he was a fisherman and we were rather keen on cormorants. 'That's it!' he said, but then waxed lyrical about a sea-eagle he'd seen on Mull on a fishing trip.

Little auks – and little belief

One November day there was a massive westward movement of thrushes visible from my office window at The Lodge. All day, flocks of redwings and fieldfares headed west at tree-top height. I kept glancing out of the window to check them out. I was on the 'phone to a colleague in Edinburgh when I saw a flock of about half a dozen birds circle the meadow in front of The Lodge – they clearly weren't thrushes and as I reached for my binoculars with one hand whilst holding the 'phone with the other I thought they might be teal. The binoculars revealed that they were little auks – presumably confused and swept up by the flocks of thrushes. It transpired that that day there were hundreds or thousands of little auks on the Norfolk coast and that many were seen at reservoirs inland – but I was probably the only person to see them over a wood. Nobody believes me to this day.

Just a quick cuppa

In Morocco, out in the Sous Massa National Park after several days charming civil servants in my peculiar French, we were *en route* to see most of the world's bald ibis come to roost when we stopped at a village. One of the project staff, Mohammed, lived there and was keen that we visit his house and have tea. As we went into the courtyard the girls of the family, all veiled, ran giggling away (and then peeked out at us from behind a door). We were shown a couple of cows and it seemed right to praise them as very nice cows, and so we did. A large pile of bright green vegetation lay on the dusty floor – cow food, no doubt. We were led into a simple room where we sat on the ground on cushions and watched our host prepare the mint tea using fresh mint leaves, a block of sugar, black tea and water. There was lots of pouring of tea from pot to cup and back into the teapot. Bread was brought and we dipped it in honey. It was an incredibly calming moment – receiving hospitality in a foreign country, from a man I hardly knew, conversing in broken language but feeling utterly happy as we nodded and smiled in true communication. Then, outside in the courtyard, in the sun, a male Moussier's redstart alighted on that pile of fresh green vegetation. My first (and only). True happiness.

That was cool

In my first year as Head of Conservation Science it was nose to the grindstone time. Even to the point that I didn't go to the races at Cheltenham for the Festival in March. Back in 1992 communications were simpler and it was indeed easier not to know the result of a football match or horse race than it is now. So I had set the video and was driving home hoping to watch the race. And indeed hoping that a horse called *Cool Ground* had done well. I had won money on this horse before – when he won the 1990 Welsh National, I had bought myself a 'new' overcoat from the Oxfam shop opposite Corals in Weston-super-Mare. There were eight horses in the race with a stand-out favourite called *Carvill's Hill* who was a monster of a horse. As I reached the Sandy roundabout at 6.25 p.m. I heard the Radio 4 news say 'In today's Cheltenham Gold Cup an outside…' before I switched the radio off. What were they going to say? An outsider won? An outsider fell and brought down the favourite? What? *Cool Ground* was an outsider and my £5 each way was down at 50/1. It was a great race - not for *Carvill's Hill* who came fifth of eight - but certainly for the new Head of Conservation Science who pocketed £312.50 of winnings.

April finches

I was walking down the Euston Road when the BBC phoned – it was probably about 25 March. They wanted to do an April Fool broadcast about birds on the Radio 4 *Today* Programme – would we play? Of course we would! I can't quite remember their proposal but it involved bird song and national anthems and I didn't like it much. Instead we came up with a storyline that followed an interview we had done a couple of weeks previously about it being a good winter for finches, such as redpolls, appearing in people's gardens. Our hoax was that too many of these finches were staying in the UK and we were worried they hadn't flown back to Scandinavia. We might be calling round to your house to catch your birds and they would be shipped out through the new, and scandalously 'not working very well' Heathrow Terminal 5. Tom Fielden came out to The Lodge on a Saturday and Jeff Knott and I did the interview. When broadcast most people got it, most of them thought it funny but a few 'were not amused'. That's humour I guess.

Whose sparrows are they anyway?

In Washington DC I stepped out of an office onto a pavement (actually a sidewalk, I guess). A small flock of house sparrows flew up in front of me. How amazing – an experience that you simply can't get in London these days but can in the USA, where house sparrows are non-native.

Foot and Mouth

When foot and mouth disease broke out in the UK in 2001 there was much alarm about whether people travelling in the countryside would spread the disease from farm to farm. We were concerned about our nature reserves, where there are lots of cattle which are mostly owned by others – not by us. After a bit of thought my team decided we should close our nature reserves to visitors to limit any risk. I stepped into Graham Wynne's office, told him what I thought we should do, he agreed and we did it. Our reserves remained closed for months costing the RSPB quite a lot of money but it felt like the right thing to do at the time and it still does. This was a very tangible indication of the RSPB's support for British farming which, to be fair, was recognised by the farming community at the time and we were thanked for it.

Cleaning up

North of Agadir in Morocco, Chris Bowden and I were watching a bald ibis pecking and probing excitedly at a very small patch of ground. The bird actually looked excited – and that takes some doing for a bald ibis. It must have stayed on the same patch for ten minutes until some children happened to scare it away. We went to investigate what special habitat had kept it so happy for so long. It was a dead dog crawling with juicy maggots. Ibis conservation solved – put out some dead curs!

Islay

The Oa is a peninsula on Islay where, on a boyhood family holiday in the early 1970s, I saw my first ever chough. On a working visit in October 2007, we saw a magnificent golden eagle, resembling Tennyson's poem ('he clasped the crag with crooked hands, fast by the azure sea he stands') and then on a Council visit in 2010 we saw a chough in just the same spot. Wonderful! And it is now an RSPB nature reserve. And as we stood by the vans on the last visit Jerry Wilson spotted a dotterel running around in the field next to us – magic.

A slap in the face

I was on leave, standing in the car park at Cheltenham racecourse on the second day of the 2009 Festival Meeting, when my phone rang. It took me a while to recognise the member of staff who was calling as he sounded worried and upset. He was warning me that he might appear in the media and thought he ought to warn the RSPB since it might be a high profile story. My mind raced. A sex scandal? Was he a closet egg collector unmasked? What? The previous evening he'd been at Old Trafford when 'the chosen one's' Inter Milan had lost to Manchester United. Jose Mourinho had not taken the defeat well, or at least the reaction of Manchester United fans, and had lashed out and hit my colleague. The story was in the Sun the next day with a witness account by two other fans. The Sun's account said that the fan involved was 'in his forties from the Cambridge area'. Yep – he was. Looking back at the results – I don't think I backed a winner that day.

No-one else noticed

If you spend a lot of time in cities, it's quite a common experience to be looking at some interesting wildlife while surrounded by people who are oblivious.

Once I was in the queue to get into the Palace of Westminster watching four peregrines in the air together. Clearly a pair and their two fledged young. Then I noticed a young falcon dismembering what I guess was a pigeon on the roof of Westminster Hall. No-one else gave it even a look. But it is thrilling that peregrines are now such a part of our inner cities. Leonardo de Caprio stopped at our peregrine viewing site at Tate Modern once.

Harry and shooting

Around the time that Prince Harry was being questioned by police about any role he might have played in the shooting of two hen harriers at the Sandringham Estate, there was also speculation on another matter – would football manager Harry Redknapp move to Newcastle to become their manager? The *Guardian* back-page headline read, 'Don't send Harry north for shooting practice'.

Car crash

I once gave a talk to a renewable energy conference where I said that the RSPB was keen on renewable energy such as windfarms, provided that they were put in the right places. I then spelled out our objections to the proposed London Array in the Thames Estuary as this area was very important for red-throated divers and other wintering waterfowl. A friend sitting in the audience heard one man in front of him turn to another and comment: 'We should organise a car accident for this guy'. I always remembered this when the RSPB was sometimes accused of being in the pockets of windfarm developers.

Being partisan

I worked with Prof. Sir John Burnett when we were both trustees of something called the National Biodiversity Network. Sir John had been in the Special Boat Service during the Second World War fighting with the Partisans in Yugoslavia. The elderly John Burnett told me the story of when the young John Burnett had been sitting in a cave one evening reading a book on the flora of the country when a Partisan noticed the book and asked why it had been written by an Englishmen. After some discussion the Partisan opined that after the war had been won all the books on Yugoslavia's wildlife would be written by Yugoslavs. That Partisan became the first Yugoslav President after the war – President Tito.

Elliot and aquatics

I once took the former environment Minister Elliot Morley to Belarus to see the results of a Darwin Award-funded project on peatland restoration. After an interesting day seeing the work (and great snipe, great spotted eagle, white-winged black tern, etc., etc.) we were given a marsh-side evening picnic by the local bureaucrats. The food was simple, but good – fish and meat and bread – but the meal was dominated by vodka. Protocol demanded that everyone at the table made a toast and that a shot was downed for each one. There were 13 of us! Elliot made a long toast about fraternity across Europe which went down very well in this ex-Communist country. It was a lovely setting, and if it weren't for the unavoidable loss of brain cells would have been an unforgettable meal – and all accompanied by the songs of aquatic warblers in the adjacent marsh.

Gordon Brown

Before the meeting of world leaders to discuss climate change in Copenhagen in December 2009, environmental NGOs organised a march through the streets of London – called The Wave. It took me back to my student days of marching against apartheid and the National Front, but this time, after the march, I was one of a small group of people invited into Downing Street to talk to the Prime Minister, Gordon Brown. We sat around the Cabinet table drinking tea. Mr Brown listened politely – he seemed quite relaxed. It was nice of him to give us the time. I wrote in my blog:

> *The Prime Minister sketched out how he saw the landscape before Copenhagen – and he demonstrated a good understanding and engagement with the issues. It can't be easy being PM – you can't know everything and be interested in everything, but Gordon Brown is clearly engaged with climate change and the task ahead in Copenhagen. He said he wanted to get a good legal agreement within months of the Copenhagen meeting and that Copenhagen should mean that countries who have offered nothing should come up with something and those that have offered something should offer more. He was speaking to other Prime Ministers and trying to get agreements before the Copenhagen conference.*
>
> *You can't disagree with the analysis – and I said so. But I asked*

the Prime Minister to realise that he wasn't going to Denmark to help get a deal – but help get a solution to a real physical problem. Agreement would be good – but a good agreement is needed. And I said that we could currently be proud of what the UK has done, and the leadership that the UK has shown – please do your best Prime Minister to come back from Copenhagen with more reasons for us to feel proud of the UK leadership.

Copenhagen was not a success.

Say it with flowers

Environment Minister Jim Knight was in India at the same time as I in February 2006 and I persuaded him to come and see our vulture project north of Delhi at Pinjore. A welcoming party of local dignitaries was organised and I was chosen to be part of it. The ministerial car drew up and one by one we greeted Mr Knight with a bunch of flowers. He looked surprised to see me grinning at him and said he wasn't used to getting bouquets from the RSPB. But he was the perfect minister – showed genuine interest, talked with enthusiasm, asked intelligent questions and smiled a lot to the Indian hosts. Mr Knight, now Lord Knight, left us all feeling buoyed up by his visit.

Huw I-D and BoP pledge

Our bird of prey pledge was signed by 210,567 people. We organised its delivery to Huw Irranca-Davis, the Environment Minister, where we assembled our group into a giant outline of a peregrine falcon in the small park by the Houses of Parliament.

Ben Bradshaw and cormorants

I was actually in Liverpool, and could see the Liver building with the two stylised cormorants on its roof, when we heard the announcement by Environment Minister Ben Bradshaw that he was allowing more cormorants to be culled (because fishermen had called for it). I was soon in the BBC radio studios recording an interview for, I think, the BBC afternoon *PM* programme. After talking about why cormorants are lovely birds (!) and how more licences weren't needed, the interviewer asked me a general question to which I replied: 'Well you have to remember that Mr Bradshaw claims to have been harassed in the street by cormorants. It's difficult to take him seriously after that!' Mr Bradshaw's remarks were quoted in the *Independent* newspaper of 28 July 2004 by Marie Woolf the political correspondent. The interviewer said: 'Well I was going to ask another question – but that's a fantastic ending!'

Red kite roost in Segovia

Back in the early 1990s the red kite was listed as a globally threatened species – probably a bit pessimistic at the time but the fortunes of this bird have ebbed and flowed several times over the last few decades. Jane Sears and I visited Spain in February 1992 and spent some time with the great Eduardo de Juana who was a university academic in Madrid and the president of SEO. We sat on a cold hillside north of Segovia one evening to count red kites coming into roost – 400 poured in! More kites in that roost on that evening than the whole UK population at the time.

Lights off in Downing Street

I was in 10 Downing Street for a meeting with Prime Minister Gordon Brown's environment advisor Michael Jacobs, and after we had chatted the group moved off out of the room. I was the last to leave and looking back saw that all the lights were on – I switched off a couple of table lights and was just turning off another when Michael came back into the room and caught me – 'Good idea' he said!

BGBW in Downing Street

One year Elliot Morley and I did the Big Garden Birdwatch at 10 Downing Street with a group of children. The list wasn't bad actually – the garden at the back of Downing Street is quite large. I think goldfinch was probably the 'best' bird.

BGBW at the US Ambassador's place

I did the 2010 Big Garden Birdwatch in the garden of the American Ambassador, Mr Susman, in Regents Park. His garden is quite large! A group of us, American and British, scoured the grounds and crunched through the snow. A fieldfare, a ring-necked parakeet and a singing wren were all good records. The hospitality of our American hosts was warm – hot chocolate for us while the Ambassador puffed on a huge cigar!

The power of the blog

I once got a phone call from a government department asking whether I could help them with a disagreement they were having with another government department. After expressing surprise I agreed, since the suggestion might well help the RSPB's advocacy objectives. I agreed that the first department could put the second department in touch with us, and in passing, said that I would probably be blogging on the subject anyway. A couple of days passed and the expected phone call didn't materialise so I asked whether we might expect a call. The reply came back that a Minister in the second department had read my blog on the subject and the disagreement had been duly resolved.

Bald eagles high

I remember sitting in a restaurant on Chesapeake Bay talking to staff from National Audubon – our BirdLife partner in the USA – as two adult bald eagles soared around outside.

Across the strand

England were playing Algeria in the 2010 World Cup on 18 June 2010. The game was a dull 0–0 draw but I was spared watching it as I was on North Uist on Vallay Island eating delicious local steak and watching hen harriers, bar-tailed godwits and little terns and listening to corncrakes. We had walked across the bay at low tide, 45 minutes across the sand, and it was a fabulous evening. A memory to savour.

The guy at the party conferences

While attending party conferences I noticed that the same man was present at all of them, usually with a placard about something (smoking, vegetarianism, self-sufficiency, etc.). After a while I got talking to him and learned he is Stuart Holmes, who once commandeered the fourth plinth in Trafalgar Square. Mr Holmes told me that he has attended the Party Conferences for over 20 years, lives on the streets and is the greenest man in Britain! He may well be! He is quite intense and treated as a bit of a nutter by many, but I admired his dedication and the way he promotes his ideas in all weathers and despite some antipathy. I told him that he would always be welcome to a few meatless sandwiches from our Fringe meetings.

David Davis

At the Conservative Party Conference in 2005 in Blackpool I was in the hall when David Davis, a leadership contender, made his speech to the party faithful. It was an interesting speech but had such a weak ending that the hall didn't realise that he had ended. After an embarrassed pause, clapping broke out. I phoned my bookmaker and put a bet on David Cameron immediately and weeks later landed my bet!

Lee Evans

In his excellent book, *Tales of a Tabloid Twitcher*, Stuart Winter describes an encounter between Lee Evans and me at the Bird Fair in 2009. Stuart lit the blue touch paper and then stepped back to watch. I mostly listened, as I recall, as Lee told me his views on white-tailed eagles, ruddy ducks and a range of other issues. I was quite impressed when he told me that he knew the location of every single ruddy duck at that very moment right across the UK. He's a very impressive guy.

Lady on 3,700

The caricature of a keen birder is a middle-aged man with a beard sitting in a hide keeping a list of species in a rather nerdy way. But there are very few people like that really. However, I was slightly surprised once during a dinner with an RSPB Local Group when a lady of a certain age sitting next to me revealed her world list was 3,700 species and she had another two globe-trotting trips planned in the next few months which would take her over the 4,000 mark.

Sleeper to Berlin

I travelled to an Anglo-German meeting in Berlin to discuss climate change and agriculture. Of the 30ish UK attendees all had flown save me, a young man from Natural England and an even younger lady from Forum for the Future. The outward journey on the sleeper was enlivened by the company of a fairly relaxed Pole and his bottle of port. Carbon reductions are not the only good reasons for train travel – as we approached Berlin there were cranes in the fields.

We are always in opposition

It's a lame joke, but I used it dozens of times at Party Conferences over the years. Whereas political parties can expect to regain power eventually, NGOs like the RSPB are never in power, always in constructive opposition! I told you it was lame – but politicians have laughed at it wryly or with greater enthusiasm for over a decade.

Forster's tern in Edinburgh

On a visit to Edinburgh for a meeting back in 1995, a group of us were down in Leith to see an overwintering Forster's tern before breakfast. That's a good way to start a day's work.

No laughing matter

The RSPB Belfast Harbour nature reserve is a great place to do some urban birding. It's a good place to see black-tailed godwits and a range of other water birds. But my strongest memory is before the reserve infrastructure of hides, etc. was in place when Dave Allen and I saw a laughing gull and a Mediterranean gull in the same binocular frame.

Robert Gillmor and RSPB Medal

At the AGM in October the RSPB gives its annual medal to someone who has made a big contribution to bird conservation. One year the recipient was the modest artist Robert Gillmor. I happened to be sitting next to Robert in the front row when he sat down after his acceptance speech. He opened the box, looked carefully at the medal with its images of avocet and other birds of conservation importance. He fingered it and I wondered what he would say. 'Ah yes' he said, 'I'd forgotten that I designed this.'

Long-eared owl at work

I don't see long-eared owls that often and so it was a delight when one arrived at The Lodge and roosted in a birch tree in view of many offices and the canteen. We saw it every day for ages – and put it on the front cover of *BIRDS* magazine.

Lodge purchase

The RSPB moved headquarters from Ecclestone Place near Victoria, to The Lodge in 1961. One story goes that the Treasurer at the time, Mr Norris, who did excellent early work on corncrake declines, was told to find a new location. He returned to RSPB Council to say that he had found somewhere he thought suitable – and had bought it! But not to worry, if Council didn't approve he would keep it himself! They did approve and for what seems the amazingly low sum of £35k the RSPB had a new headquarters. We also believe that Princess Margaret was interested in the property but that the proximity of a right of way to the main building put her off. How different life could have been.

Break in

RSPB staff have many and various skills. I once watched with admiration when Jane Sears broke into a locked car in double-quick time with a coat hanger – our Spanish host was both relieved and amazed at her dexterity.

Hand of God

In 1986 we took a day off from surveying so that we could watch England play Argentina in the World Cup – Scotland weren't involved! Most football supporters can remember where they were when Maradona cheated to put the first goal in and then scored one of the best World Cup goals ever a few minutes later. I was in a static caravan at Lairg in Sutherland.

Greylag in the Flows

Walking across the Flow Country one overcast day I disturbed a greylag goose from her nest on a small clump of vegetation in a small lochan. The goose flew off calling and circled around me at a distance of a hundred metres or so, high up and still calling. For a moment it felt as though she and I were the only living creatures on Earth. I stood and looked at her, she circled me and called. I shivered, walked on and the sun came out.

Bomb

On 13 October 2010 while working at The Lodge, the tannoy told us all to go home. We later learned that an old World War Two bomb had been found. I wrote in my blog a few days later:

> *After we had gone, a World War Two bomb was safely removed by bomb disposal experts. They were pretty sure, 95%, that it would not have exploded but, hey, 5% is 5%! When the news of a suspect device circulated, speculation mounted about whether the Provisional NFU were active in the area.*

Action!

I like the film *Gladiator*, and its opening scenes ('Unleash Hell'!) are very striking. But I'm always distracted by the fact that the woodland in which it is set is actually right next to the RSPB's Farnham Heath nature reserve. Also, a short scene from the film *Atonement* (unfortunately not including Keira Knightley) was set on our Nene Washes nature reserve. Nice little earners.

Not enough fish in the sea

Back in the late 1980s we were worried about over fishing of sandeels by the Shetland fishermen – seabirds were starving to death. We tried lots of ways of putting the case for reduced fishing pressure but what seemed to work best was the publication of an article on the subject in *New Scientist*. We were told that it was seen by the Secretary of State for Scotland in Edinburgh, Ian Laing. He asked what on earth was going on and fishing restrictions quickly followed. How to close a fishery in 1,200 words?

Bird Brain of Britain

At the annual Bird Fair at Rutland Water there is a competition, based on *Mastermind*, where competitors have a round of specialist questions and then a general knowledge round. I was once the RSPB entry and my specialist subject was Globally Threatened Birds of Europe. I didn't do very well and think I was last after the first round, but stormed ahead to win in the general knowledge round. This was a big relief considering the RSPB's excellent record to that date and the fact that RSPB vice-President Bill Oddie was the eccentric question master. Years later David Lindo, the Urban Birder, reminded me that I had beaten him – he was working for the BTO at the time. Never mind David – it didn't do my career that much good and it hasn't held back yours.

This firm has spunk?

Sign seen over a Ghanaian garage – Black sperm motors – why?

A wildlife spectacle – or a freezing cold morning?

My colleagues from the marketing department organised an October, early-morning, high-tide visit to Snettisham for potential corporate donors which illustrated the unpredictability of taking people to a nature reserve to impress them. We'd spent the evening before getting to know each other at a local hotel and it was a bit of struggle to get out of bed early next morning and head off down the shores of the wash to Snettisham where another cold walk through the wind brought us to the nature reserve. In the next hour, around a quarter of a million knots flew over our heads and started roosting on the shingle bank as they fled the incoming tide. I loved the sight of it and tried to share my enthusiasm with the group. I'd say about half of them loved the experience and half hated it. With half we probably increased our chances of funding, but with the other half we dramatically decreased it. Some were just left cold by the whole experience.

Taking your opportunities

The 2008 Conservative Party conference was in Birmingham so we took the opportunity to invite the Shadow Secretary of State, Peter Ainsworth, to our local nature reserve at Sandwell Valley. This gave me an opportunity to lobby Peter on various issues and him the opportunity to flee the party conference for a few hours. Peter was a delightful visitor, and has a true interest in nature (he now chairs the excellent NGO Plantlife), and I was sorry that he was shuffled out of the shadow cabinet by David Cameron the next spring. However, our visit ensured that nature conservation, Sandwell Valley and the RSPB got decent mentions in Peter's conference speech the next day.

Bird is the word

In the run up to Christmas 2010, the RSPB produced a video to promote an annoyingly catchy song, *Bird is the Word*, in its bid to be Christmas Number One. I was asked to do a little cameo in the video which largely featured a whole host of talented RSPB staff dancing in the snow on the lawn outside The Lodge. *Bird is the Word* got to Number Three and I still get comments of 'nice moves' occasionally. It was fun, took an hour of my time and was a way of showing a different side of the RSPB – and can still be seen on a certain popular video streaming service!

Harlequins

We stayed in Ayr on a trip to southwest Scotland one winter to talk to regional colleagues. An early morning start enabled a quick trip down the coast to see three long-staying harlequin ducks bobbing just off shore. We were back in Ayr for breakfast and then the work went on – it's great to work in an organisation where the love of birds runs through all that one does. I sometimes thought of that morning in my Lodge office when it became infested with the recently arrived harlequin ladybird.

Cleaned up

In Birmingham for the Conservative Party Conference in 2010, we took the opportunity to set up a media thank you event for John Struczinski, a cleaner at Birmingham Airport. His observational skills led to the capture of Jeffrey Lendrum. John alerted security when he noticed that the floor was not wet after Lendrum left the shower and there were discarded egg boxes lying about. When searched, Lendrum was found to have 14 packages strapped to his body. Explosives? Drugs? No – peregrine eggs.

Heat

While we were campaigning against the daft idea to build an airport at Cliffe, an anti-airport march was organised in London from Parliament Square to Trafalgar Square. Jamie Oliver was at the front of the march with his wife and young baby, as his parents lived near Stansted. I realised that the young chef was going to be the focus of all the attention and so stuck behind him with a 'No Airport at Cliffe' placard. Almost every photograph of the event in the Sunday papers had that placard in it. Later in the week I was shown a copy of *Heat* magazine which had a similar picture of the Olivers with me standing behind them – the caption mentioned the man with the enormous moustache.

Counting sheep

Driving north down Strath Halladale, returning from a long day's bird surveying, the road passes an outcrop on the right. I was amazed to see a sheep come tumbling out of the sky, falling towards the road. I braked soon enough to stop in front of the very fluffy sheep as it hit the road surface. Its unshorn condition probably saved its life as it landed on its back which was well-cushioned with tangled wooliness. It was shaken, but wandered off after we helped right it. We were a bit shaken too – a fraction of a second earlier and that sheep would have landed on our bonnet or roof and that would have been a bit of a shock to the system out of the blue. And how would you rate your chances of persuading the RSPB's Transport Office that the dent in the car was because a sheep fell on it?

First radio interview

I've done a lot of radio interviews in my life but the first was as a schoolboy at Bristol Grammar School in the early days of BBC Radio Bristol, in autumn 1970, with local bird man Robin Prytherch. The short interview was one in a series about children's hobbies and I and two friends talked about…yes! Birdwatching.

WWT competition

The Wildfowl Trust at Slimbridge used to have an annual competition for local schools in three age groups. We at Bristol Grammar School always used to enter several teams and sometimes we would win. But the most winning school was Leighton Park near Reading. I mentioned this while at Slimbridge recently and Nigel Jarret found the old trophies stored away. Of course winning was a thrill but it was all the sweeter because the trophies were presented to us eager schoolboys by the great Peter Scott himself.

Keith Brockie saved my life

We arrived on the Hardangervidda plateau in early June 1978 – long before all the snow had melted and the lakes were free of ice. One day Keith Brockie and I were crossing a frozen lake when, despite being rather sylph-like in those days, I disappeared through the ice into the freezing water below. Keith Brockie helped me out – and without his assistance I think I would have been a goner. Youthful lack of care.

Netting for bats

Whilst studying pipistrelle bats I got permission from many churches to catch bats as they left the church at dusk. This was often over the west door or sometimes the main church entrance. One evening I arrived at the church at Kingston, put up my mistnets and sat quietly in the churchyard listening to the rooks and jackdaws. I checked the net every now and again but it seemed like a quiet night – which was a shame as this was a church where I had caught a Natterer's bat on a previous occasion. I was about to pack up and leave when the church door opened and I captured the exiting parish council meeting in my net. After a bit of confusion there were good-natured laughs all round.

Vicar looking like Dick Emery

Older readers might remember the Dick Emery comedy TV show. A variety of characters appeared every week including a buck-toothed vicar. When studying bats I knocked on the door of the rectory at Wilburton in The Fens to ask permission to study bats in the church and a Dick Emery look-alike opened the door. I asked permission to have a look for bats and mentioned that I used sulphur smoke to encourage the bats to leave their crevices. The vicar thought for a while and said it was fine, and if his congregation ever remarked on it he would point out to them that there were only two smells in the next life – brimstone or incense – and that most of them had better get used to the brimstone.

Rose-coloured starlings

There was a time when there was one particular bird that I had seen, but not through my aged pair of Zeiss 10×40b binoculars. While sitting in a bee-eater colony in the south of France one May day, I glanced up to see six adult rose-coloured starlings flying past – unmistakeable and amazing. My binoculars were right next to me but my hands were filled with a bee-eater and a syringe with which I was taking a blood sample. It was about ten years until I caught up with another adult – in Northamptonshire, with bins to hand.

Cost of oil

During my PhD study I drove around East Anglia in an old Austin 1100 car. It was a constant drain on my limited resources with road tax, MOT certificate, new tyres and servicing. I could claim some expenses from the Natural Environment Research Council, who funded my study, but they questioned one claim for oil for the car as being on the steep side. After I had pointed out that had I studied distant galaxies they would have provided a radio telescope (I wouldn't have had to buy a second hand one myself) and if I had been studying cell membranes they wouldn't have made me buy my own electron microscope the money came through.

100 km from Calais

On our way back from the Camargue one year, John Krebs and I were on the autoroute north of Paris and had just passed a sign declaring 'Calais 100 km'. We promptly cheered – and the engine of our van blew up and left us stranded in northern France. The van was towed to a local garage in Bully les Mines, a depressing mining town (or maybe it was just I who was depressed) and I raised my spirits by playing Space Invaders in the station café as we waited for a slow train to the coast.

Piles of money

I stayed in a hotel in Accra, Ghana, for nearly a week and the bill had to be paid in cash in the local currency, the Cedi. Cedis came in low denomination notes and a trip to cash a few travellers' cheques had me carrying the money back to the hotel in a bag. I thought I knew how much the bill would come to so I counted out the greasy notes in my simple bedroom. It was a substantial pile of paper money – about a foot high. So I spent quite some time counting and recounting a pile which I thought would cover the bill. The next morning when I came to pay I put the pile of notes, a foot high, on the counter. The man on the other side looked at it, tilted his head and looked at it again, turned his head another way and looked once more – and then put the notes in the safe without counting them at all.

Bees

Working on bee-eaters meant that I had to show a bit of interest in bees, and that meant bee-keepers too. One day in France a bee-keeper came to look at some neglected bee-hives and for some reason I was roped into 'helping'. We wandered down to the hives and he got out a bunch of old newspaper which he stuffed into his smoke-blowing contraption and then set it alight. A few puffs of smoke seemed to please the bee-keeper and he gave me a bag to hold while he quickly puffed some smoke into one of the hives, whipped off the roof and pulled out one of the slats covered with bees and honey to inspect it. This all happened very quickly and was fascinating until I realised I was standing in a T-shirt and shorts, with my hands full, next to a beehive, with the surrounding air full of thousands of bees. The bee-keeper was engrossed in his task and seemed unconcerned so I tried to be unconcerned too. I must have stood there for a few minutes and only collected a few stings on the legs, but it was quite an experience. When I tell this story to bee-keepers they gasp and say that we were both completely mad.

Central lobbyist

I've often sat in the Central Lobby of the Palace of Westminster waiting to meet an MP or Minister. It's an amazing crossroads within the fine architecture of Pugin's Palace. Above each of the four passage ways stands one of the UK nations' patron saints. St George, a warrior, faces the House of Commons where politics is at its most embattled and adversarial; whereas the orator, St David, faces the House of Lords where thoughtful and polite eloquence is more the order of the day. St Andrew is looking towards the nearest exit and St Patrick towards the nearest bar. But as one sits there you'll see familiar faces from the TV pass by on their busy ways.

Say it with roses

We were finding it difficult to get a meeting with a politician's Special Advisor so I decided to try a new route. I sent him a red rose through *Interflora* with a note saying 'Don't be a stranger. We should talk. The RSPB' and we got a meeting very soon afterwards.

Swallow on the wire

My regular drive into work from east Northants to Sandy was through an arable-dominated landscape with only one farm on the route still primarily dealing in livestock. This dairy farm, where sometimes I had to wait for the milked cows to cross the road, was almost always the first place away from the local gravel pits where I would see a swallow each year. Sitting on the wires above the cow shed, feeding on the insects that gathered there, the swallow, with its long tail streamers, was always a sign of spring for me.

Lapwings and Leo

I was giving evidence to a parliamentary committee on 20 May 2000, on the subject of farmland birds, just as the Prime Minister and Cherie Blair welcomed the birth of their fourth child, Leo. I made a point about the decline in lapwing numbers and that if his Dad didn't do more to fix farmland bird declines then Leo would grow up in a country where there were no lapwings left in the countryside. There was a mixture of smiles and tutting from the committee at this off-the-cuff remark, but it found its way onto the radio programmes *Today in Parliament* and *Yesterday in Parliament* as a topical parliamentary reference and thus got the plight of farmland birds, and the name of the RSPB, into the public and political spotlight for a few seconds.

Grouse meal

I spent a night at the home of a grouse moor owner in Yorkshire. He and I got on well and there was quite a lot of good-natured banter during my stay. At the evening meal, to which he had invited some friends, my host was clearly delighted to place a plate of grouse before me. Everyone looked at me with smiles of anticipation – I don't know quite what they expected, I'd eaten grouse before – but I looked down at my bird and said: 'Thank you – if it's good enough for hen harriers then it's good enough for me!'

Political grouse

I was once visited at The Lodge by a grouse moor manager who proceeded to tell me that he was a very rich and powerful person with very close links with the Conservative Party and that he would make sure that when the Conservatives got back into power, as they surely would, the RSPB would have no access to Ministers at all. This was a threat that took quite a while to be tested, as the Labour Party won the next two general elections. But it hasn't proved to have much substance since – an organisation with over a million members can't easily be ignored by any government.

The price of everything

A grouse moor manager came to The Lodge one sunny day and, as was my normal practice, I took him for a walk around the attractive grounds after lunch in the canteen. I used to do this with visitors partly so that they would see that this was quite a big operation with c. 500 staff working at the headquarters. I asked him to explain to me what he saw in grouse shooting and he started by saying that it was like anything in life, you have to pay through the nose for the real pleasures. He lost me there. The real pleasures in life – love, friendship, a sunrise and the beauty of nature for example, can't be bought. We were starting from very different places.

Medal

At the RSPB AGM in 2010 we gave Prince Charles the RSPB's highest honour – the RSPB Medal. He was, many thought, a most deserving recipient and the 800-strong audience seemed delighted that he was to receive it. HRH was not present personally – in fact he was opening the Commonwealth Games in Delhi the following day – but his acceptance speech was read out. After praising various aspects of the RSPB's work the Prince moved on to one of his hobby horses – grey squirrels. He praised our work on red squirrels but his parting shot (should I use that phrase?) was to hope that our good work on grey squirrel could be transferred to other species whose 'burgeoning populations have become a menace in this crowded island'. What did he mean? There were more than a few of us who believed that the Prince had just narrowly avoided slagging off birds of prey.

Greek tragedy

I've been introduced to Prince Philip only the once, but it was a brief and chilly meeting which left others puzzled as to the Prince's behaviour. I was at an event graced by his presence, and groups of people were collected up to be introduced. One by one, individuals were presented to the Prince who would smile, shake hands and ask a question or two before being introduced to the next person. When my turn came there was no smile – just an icy silent stare. I met the Prince's gaze and we looked each other in the eye, neither breaking gaze. He then turned to the next person in our little semi-circle, smiled, shook hands and said something. When he moved off several people asked me what that was all about and I simply said: 'Sandringham and hen harriers'. Realisation spread across their faces.

WSSD

During a conversation in a pub, Andrew Balmford, Rhys Green and I hatched a plan to produce an economic analysis of the value of wildlife to humankind. It was a long range plan as we aimed to produce it in time for an international meeting in Johannesburg, nearly a year later. All went to plan and a clutch of experts assembled in Cambridge in early 2001 for a week of thought and work. When they left I wasn't sure that the group would hold together but they did, and they produced a paper which was accepted in the American journal *Science* to be published to coincide with the South African meeting. Rarely do such plans come together so well. There were plenty of potential problems *en route* but we had avoided them all and were set up for a headline grabbing news feast at just the right time. Unfortunately, in the same issue of *Science*, there was a more interesting story – about a jackdaw using a bit of bent wire as a tool. To add insult to injury that paper was by a former colleague of mine from Oxford, Prof. Alex Kacelnik.

Ragwhat?

We once had a visit to Hope Farm from several luminaries of the Country

Land and Business Association. It was a good meeting and we all walked around the farm and looked at the crops and skylark patches. The CLA regard ragwort as a terrible weed and campaigned against it then, as now. We stopped at one point to talk and one of their number said, 'You've got an awful lot of ragwort here' and his companions all agreed with him and each pointed to a different nearby yellow flower. Amazingly, not one of them was actually a ragwort plant. Strong opinions on wildlife problems are not always based on great knowledge.

What's that swan?

I'm not a bad birdwatcher but there are many much more skilled and able than I and so I tend to remember small triumphs (as I have very few large ones). I was once on Orkney with the RSPB Orkney Officer and expert-birder Eric Meek, looking down at a swan which had upended in a lake (actually, I guess it was a loch). It seemed to be able to hold its breath for ever and Eric wondered whether it was a whooper or a mute and said we'd have to wait to see its head. 'It's a whooper' I said, before this was confirmed a few moments later by the bird revealing its yellow and black bill. Eric asked how I knew or whether it was a lucky guess and I pointed out that the swan's tail was rounded rather than pointed like a mute's. Eric nodded in a way that I interpreted as,'Maybe some of these people who live in offices do know a bit about birds'. Years later I was reminded of this when a dead swan was found in a harbour in Fife with the H5N1 bird flu virus in its corpse. It was first thought to be a mute but days later was confirmed as a whooper. The news of the corrected identification reached me by text message in a pub in York, at the RSPB Members' weekend in April 2006, whilst settling down to watch the Grand National which was won by *Numbersixvalverde* at 11/1, with none of my money on it.

Just a name

I probably pronounce my surname more carefully than most do as I am accustomed to being called Aviary and then told that it's a very suitable name for someone with an interest in birds. I'd say about six people per year tell me this and I try to look shocked and amazed for every one of them.

> *If history were taught in the form of stories, it would never be forgotten.*
> **Rudyard Kipling**

Chapter 15

Whither the RSPB?

Strategy without tactics is the slowest route to victory. Tactics without strategy is the noise before defeat.
Sun Tzu

I worked at the RSPB from 1 April 1986, the day I first headed off to the Flow Country bogs, until the 30 April 2011 when I stepped down from the post of Conservation Director. It's difficult, after all that time, not to talk of the RSPB as 'us' and 'we' even when thinking of the future – although as a life member of the RSPB I have paid for an everlasting stake in the organisation anyway!

My curriculum vitae for the last 25 years reads rather dully as 'worked at the RSPB' but those years were a fantastic personal journey. I arrived as a scientist, was transformed into a half-decent manager (whilst not losing all my science in the process) and then became a leader, communicator and advocate (whilst still doing the essentials of managing a large budget and numerous staff, and still retaining the vestiges of some scientific understanding). At least – that's my story and I'm sticking to it.

But while I was growing (mentally, but also in weight) and changing, so was the RSPB. The membership reached 400,000 in the year I arrived and 11 years later passed the 1 million mark on its way to the current 1.1million. The 100th RSPB nature reserve was acquired in 1984 and I was very much involved in the purchase of the 200th, Sutton Fen, in 2007. International work was very much focussed on UK birds which migrated to Africa and our close colleagues in Europe, but now encompasses rainforests in Indonesia and vultures in India. Internally the network of regional staff has grown enormously, but so has the number of marketing and computing staff at headquarters. The RSPB once concentrated on rare UK birds but now involves itself in everything from house sparrows to spoon-billed sandpipers. The RSPB used to leave most of the counting and monitoring of birds to the BTO and WWT. But now it is intimately involved in funding and fostering such work itself and its data gathering and handling expertise has increasingly been used abroad.

I wonder what a scientist, warden, press officer or advocate arriving at the

RSPB this year, and working there for 25 years, might be able to look back on in 2037? That retrospective will be determined to a great extent by the decisions made over the next few years amidst difficult economic times.

All organisations must look to the future and these are some of the questions that I think the RSPB will grapple with over the next few years. It will be the decisions that its leadership makes, and then how they are implemented, as well as the buffeting of the outside world, which will determine the sort of RSPB that is handed on to the next generation. It's a precious organisation and so the decisions will be important ones.

I would claim that the RSPB was the best nature conservation organisation in the UK when I joined the staff and still was when I left. In many ways it is the most effective wildlife NGO and therefore the tactical and strategic decisions it makes in the future will have far-reaching impacts on how nature fares in the years ahead.

Birds or all wildlife?

As the RSPB moved away from bird protection to bird conservation it became adept at using the full conservation toolkit. And, once you are involved in managing land and arguing for changes to public policies then your work has much wider relevance to wildlife as a whole. Using and improving the site protection mechanism of SSSIs, SPAs and SACs does not just help conserve birds, it helps conserve all wildlife. So while it is easy to argue that the RSPB is now involved in far more than bird conservation, to what extent is it a fully rounded nature conservation organisation – and to what extent should it be?

Saying that you start with birds but that your work helps other forms of wildlife as well is not the same as saying that you start with all wildlife and work out the best things to do as a result. The extent to which the RSPB should or could become a fully rounded all-nature conservation organisation depends not only on what it wants to do but also on what other nature conservation organisations want to do and how they all fit together. That is the subject for the next chapter.

The English question

I am a Bristolian with Welsh and English parents and I live, now, in east Northants. I have a degree from an English university and another from a Scottish one. I have worked in Scotland as a nature reserve warden, a field worker and a scientist. But then I've worked in the south of France for quite a long time and have studied wildlife in Canada, Norway and Spain too. And I've visited a few dozen countries on five continents for work or holidays. The music I listen to is more likely to be German, Austrian or Italian than British, if it is classical, and more likely to be American than British, if it is popular. The authors I read are mostly English, but Americans come second and then the Irish, Scots and a rag-bag of Europeans get a look in. So what am I in cultural terms and what are my roots? At various times I can be Bristolian (though I don't have the right accent), English, British, European or a world citizen – and my love of wildlife certainly leads me to take a broader and more expansive view because wildlife is universal and wild creatures, particularly birds, move about oblivious to the lines that we draw on maps.

But within the lines on maps, my closest links are with England and an English government in Westminster and that's a consequence of where I live rather than where I was born. In a devolved UK it's the English political system that rules my life, just as it would be the Scottish one if I lived in Scotland. Devolution has made our lives in different UK countries not just culturally different but politically different.

The RSPB is a very rare beast now in being a UK organisation – one single organisation with one budget, one board of staff directors and one Council of trustees. There aren't many organisations like that these days, particularly among those that deal with the political world. The RSPB finds the new geography a little difficult in that it has Welsh, Scottish and Northern Irish Headquarters but no English Headquarters – just a UK Headquarters, sited in England. It has an operations director for Scotland, another for international work and another for England, Wales and Northern Ireland. RSPB's Council has advisory committees for Northern Ireland, Scotland and Wales but not for England or international work. Political geography is tricky.

As time goes on, it will be more and more difficult to maintain coherence between what RSPB staff are doing in different parts of the UK as local national circumstances pull the work in different directions. Just as the UK Agriculture Minister (whose day-job involves dealing with England) has to take into account the views and needs of Northern Ireland, Scotland and Wales, any policy officer working in the RSPB needs to be very clear about whether their job is European, UK or national – or even more challengingly, *when* it is each of those things according to the matter at hand.

Some of these issues would be forced if any part of the UK goes it alone – and Scotland looks the most likely (although perhaps not very likely) to do so at the moment. Imagine the impacts of Scottish independence on a UK RSPB. Would RSPB Scotland contract England/Wales/Northern Ireland RSPB to carry on storing its membership records or would that whole job pass to the new RSPB Scotland? Who would pay to set up the new system – the small Scottish organisation or the larger non-Scottish one? Would the non-Scottish organisation still be able to keep the names and addresses of the RSPB's members resident in Scotland and ask them for money? Scotland holds the greater part of the RSPB's nature reserves by area. Ownership of these nature reserves would pass to RSPB Scotland and so, presumably, would the costs of their management. Suddenly the non-Scottish RSPB would be bereft of Abernethy, Forsinard, Mersehead, Hoy, Balranald, Mousa and a host of other sites, but it would also be relieved of the significant costs of running them – costs that would pass to a much smaller RSPB Scotland membership. How would RSPB members relate to the new RSPBs – how many would join both organisations and how many would leave the one under whose wing they might be expected to stay? How should the non-Scottish RSPB regard its RSPB Scotland neighbour in the overall BirdLife International family? Would Scotland get the financial nod ahead of Spain, India, Poland or Indonesia? Maybe it will never happen, but if it does then I am rather glad I won't have to work out the consequences.

But for now the RSPB faces the challenge of being a UK organisation, without looking like an English one, despite the facts that it's global (!) headquarters are in England and, just like the UK population as a whole, around 80% of its members live in England.

UK or International?

'Think global and act local' is a phrase that has been around for over 40 years and each time I hear it I think, 'Why not think global and act global too?' The balance between the RSPB's work in the UK and its international work has gradually shifted over time towards the international dimension – but since the RSPB has grown over the same period it hasn't meant that international growth has been at the expense of growing its UK capability as well.

However, it is a very real question as to how much of its resources the RSPB should put into UK issues and how much should be devoted to different aspects of its international work – e.g. flyways that hold 'our' birds, Europe and the EU (similar, but not quite the same things), and truly international issues such as Sumatra's rainforests.

Many of 'our' birds go 'abroad' at other times of year and so bird conservation can only be done properly if action is taken in every country within a migratory species' range. The roseate tern is a case in point and was the main reason that the RSPB started working in Ghana. Roseate terns arrive on their breeding grounds in May each year, as two-year-olds or older, and have gone by September. So for eight months of each year, and for all of their non-breeding year, they are somewhere else – we believe off the coast of West Africa.

Most species are migratory to some extent. There are all those warblers and flycatchers and hirundines that arrive with us in summer having spent our winter in Africa and then all those waders, ducks, geese and swans which leave the Arctic to enjoy our mild winters. Even the robin and blackbird in your garden might nip down to France in winter to be replaced by a continental bird following a similar idea. As the seasons cycle, millions of birds come and go all the time and none can be properly conserved unless conditions are bearable everywhere on that journey.

If birds did not migrate or make hard weather movements, this first reason for international engagement would disappear. However, a second reason for an international view would remain, and for birds and non-birds alike. Political decisions that influence UK nature conservation are partly made abroad by international agreement, whether they relate to how agriculture does, or doesn't, work over most of of Europe, or how countries will, or won't, curb their greenhouse gas emissions.

The imperative to act internationally applies almost everywhere in the world, for no country is isolated in terms of political influence. But it pertains very strongly in Europe, a relatively small continent containing twenty seven EU member countries and a number of non-Union countries as well. The EU measures 4.3m km^2 – less than half the land area of the USA – and yet is divided into about half as many countries as the USA has states.

From farm subsidies to biofuel policies, and from nature directives to fishing regulations, the EU sets a common standard for a varied mix of countries – small and large, rich and poor. All of the EU states must adhere to all of those policies affecting nature continent-wide, so nature conservationists must be fully engaged with them or nature will suffer. There are swings and roundabouts in trying to influence supranational policies – you have less influence on your own but can have more impact if your influence wins through.

The third reason for working internationally is that we can help with global conservation priorities that, while they may hardly impinge on our own wildlife or habitats, need fixing all the same. If we lose the Indonesian rainforest we may lose the tigers that live there. They were never going to visit the UK and I was never going to see them in Indonesia, but it saddens me to think that they could be lost – and maybe they won't if I devote a small proportion of my conservation effort, alongside that of others including the people of Indonesia itself, to tackling that conservation issue.

There are plenty of reasons for a UK organisation to devote some resources to international nature conservation as sketched out above – 'our birds', 'our friends in the EU' and 'our planet' summarise the arguments. But while this approach expands the limits of what you can do – it opens up a great many potential ways to spend money, so there are hard budget choices to make.

Just as all domestic politicians want to be statesmen on the world stage, all conservationists are drawn towards international nature conservation. But it remains to be seen what proportion of the RSPB's primarily English membership is thrilled with the Society expanding its international work. I would hope that enough of them are keen on it to send a signal that it should gradually increase, but I doubt that many would wish for it at the expense of continued investment in new and existing nature reserves and political advocacy at home.

Where will the money come from?

The RSPB isn't rich, even though everyone seems to think that it is, although it does have considerable public support which is the main source of its financial assets. In the financial Year 2010–11 the RSPB's turnover was around £120m but once you take out the cost of income generation and money spent on goods for resale then the income comes down to a handy £94m. Just under a third of that money came from membership donations and subscriptions and another 22% from legacies – meaning that a little over half of the income comes from ordinary people like you and me (alive and dead).

We are also the people who provide the 17% of the RSPB's income that comes from commercial trading by buying Christmas cards, bird food and a large range of other merchandise. Another 5% comes from land rents and sales and returns on investments (though I could never persuade Finance Director Alan Sharpe to let me take a wad of the Society's money to the racecourse with me – perhaps because, on the one very cold day he accompanied me, I drew a complete blank!). That leaves just under a quarter of the income raised from grants, corporate sponsors, and trusts, including those subsidies that all farmers receive, grants for environmentally friendly farming and grants from grant-giving bodies that have to be won competitively against other bidders.

At the time I left the RSPB the Society had financial reserves of £42m, although all but £13.5m of these were already committed (just not spent yet), which meant enough loose cash to run the organisation for about nine weeks if all the money stopped coming in. No-one expects the money to stop coming in, altogether, all at once, but this is a hard time financially for everyone and that includes all those RSPB supporters and charitable trusts, etc. Proposals for CAP reform to cap any one landowner's single farm payments (the pure, or I would say, impure, subsidy payments) at €160,000 would hit the RSPB's income, as well as that of other major landowners.

Under these circumstances it really is vital to maintain, and if possible grow, the membership and persuade our loyal supporters somehow to pay a little bit more each year – for a million little bits more add up to a lot. I was always keen that we should grow the membership for another reason too – the more members, the more political clout the organisation has.

I've been an RSPB member for many years – initially as a YOC member and then again as an adult shortly before I joined the staff. Many RSPB members are long-term and very loyal, but not everyone is an almost-lifer like me, and, I hope, you. Some people join on a whim and then think better of it, others may be let down and not get their Christmas cards on time (it happens, but very rarely), some may just decide they are too strapped for cash this year, and yet others may resign in a rage when they hear the RSPB's Conservation Director talking about birds of prey, farming, climate change or ruddy ducks on the radio. Despite all those reasons for not renewing your membership adult renewal rates are around 90% each year which is very impressive. But let's just work it out. Ten percent of 1.1m members is 110,000 members who have to be replaced every year. That means that every day of the year the RSPB must recruit 300 new members to replace the lapsed ones – just to stand still. It's an amazing achievement and one in which I played no real role at all. Even as a staff member I stood in awe at the achievement – although I might have pointed out to my marketing colleagues that if they couldn't sell our conservation work to the public, then they couldn't sell anything!

Hidden away in these figures is the rather small contribution made to the RSPB's finances by corporate business. Many other NGOs tap business and industry for a lot more money, so there must be scope for the RSPB to get a greater share. However, it is difficult to face in too many directions at once – the public, grant giving bodies and industry as well? And industry, being businesslike, tends to want something for its investment. But this must be an area for potential future growth. Whose money should the RSPB take? Banks, oil companies, supermarkets, pesticide manufacturers, bookmakers, the media? It's a bit of a minefield but then the RSPB and other membership organisations don't limit their memberships by sieving out those with criminal records, people who work for Monsanto, Ladbrokes or British American Tobacco, politicians or industrial farmers so why, you might ask, should any companies be turned away? It's a good point but I think the difference lies in the offer. A membership organisation offers its membership to the world and anyone who takes up the offer gets the same treatment, but not their name on your masthead or posters. Working with business normally involves a visible *quid pro quo*, and taking corporate money to foster your aims somehow implies shared values in a reciprocal way quite different from taking money for an individual membership.

It's a lot easier to be principled when times are easy but in the current economic circumstances any organisation's principles tend to be tested.

Saving nature or saving us?

'Is nature conservation for nature or for people?' might seem a strange question but I don't think it has a very clear and straight forward answer. If your instincts are to say that it is for nature (let's call it Route N) then you wander down a road where you will eventually have to decide how much gain for nature is worth how much pain for people. Is a rather dull and tiny rare snail reason enough to object to a road scheme that will cut minutes off millions of journeys taken by polluting cars? Jobs or wildlife? Economic growth or a pretty countryside? Feed the world or have some skylarks?

But if your instincts tell you that nature conservation is primarily for people so that we can enjoy nature and its benefits in enriching and sustaining our lives (Route P, we'll call it) then you come to some tricky places too. Most people have managed perfectly well without corncrakes in the English countryside for almost a century so why bother to bring them back? Soil microbes are more important to the ecosystem than skylarks so that's where we should put our resources. Most globally threatened species wouldn't be missed, so why bother to save them?

I have found, and I think it was Graham Wynne who taught me this, that whenever you are faced with a tricky and uncomfortable choice it is often illuminating to shift to a more general level of argument and examine the question from that higher ground. A different way of looking at the same question is to take the advice given to Leo McGarry in *The West Wing*, 'Never accept the premise of the question'. If asked whether you are going down Route P or Route N then the right answer is often that we aren't at a fork in the road – there is a better way which involves travelling down both routes at once.

In a rich country like the UK then it isn't a choice between my standard of living and saving nature – we can do both. Indeed, given our level of scientific understanding and material wealth we can do an awful lot better at both if we put our minds to the task. We can have farmland birds back in the countryside without having to pay millions more pounds to farmers and without farmers losing noticeable amounts of income, if only we arranged things better.

Practical or game-changing?

I always used to say, and really did believe, that one of the great things about the RSPB is that it does nature conservation across the board. It is a major landowner with all the practical experience that that brings and it walks the corridors of power in the UK and abroad wielding influence to benefit wildlife. I still believe it's a combination that equips the RSPB better, in many ways, than Friends of the Earth or the National Trust. Being involved in farming on the ground makes the RSPB better informed when it comes to CAP reform and, *vice versa*, lobbying on climate change and biofuels makes the RSPB a more considered and thoughtful land manager too.

But having a foot in each of two camps is not necessarily a comfortable place to be. Can you imagine an organisation that is a combination of Friends of the Earth and the National Trust? Well it isn't exactly like that but there have been tensions in the past between how much effort the RSPB should put into practical nature conservation, like land purchase and management, reintroducing threatened species and investigations into wildlife offences, compared with those game-changing advocacy activities that will have impacts on huge areas and many species at the touch of a decision-makers pen, if only you could get them to make the right decisions.

Practical conservation work is small-scale but reliable and advocacy work is large-scale but unreliable. In which would you invest? The easy answer is that by investing in the best prospects in both areas you do a better job than sticking to just one. In racing terms, there are good bets to be had on both favourites and outsiders – the trick is spotting the value and not boxing yourself in.

But perhaps that *is* too easy an answer. Few professional gamblers making their money from betting on horse racing (to take a completely random example) would actually play across all the odds in the way I have described. Most would find that most of their money came from concentrating on betting propositions in a particular and fairly narrow range.

The effectiveness of practical conservation action and advocacy waxes and wanes with the times. Sometimes governments are listening and sometimes they are not, sometimes economic conditions favour progress in one way and sometimes in others. It is quite likely that in the current economic conditions there may be some good land bargains available and that politicians are not listening very hard to the environmental movement – so keep your hand in the advocacy arena but press on with some timely land purchase and make a difference on the ground.

What's in a name?

Given that the Society for the Protection of Birds was formed in 1891, and received its Royal Charter in 1904, there has to be some question over whether the current day RSPB has the right name. If you doubt this then just imagine that the RSPB has sprung, fully-formed, from the earth today in its current state – would you call that organisation the RSPB? I suggest that you might not.

It is often said of the British Trust for Ornithology that it is not solely British, not a trust and doesn't simply do ornithology – but apart from that the name is perfect. And a similar deconstruction of the RSPB name would reach a similar conclusion. For some years now the RSPB has rarely itself spelled out its name in full, preferring to be known by the acronym than the mnemonic.

Birds?

The RSPB is about more than birds but should that be reflected in the name or not? Do birds provide the RSPB with a unique selling point that should not be jeopardised? Is the RSPB just a club for twitchers or, in contrast, does it forget the birdwatchers, who are at its core, rather too often? Whatever the right answers to these questions, whether the RSPB has a 'B' in its name goes to the very heart of them in terms of perception, even if not in terms of reality. Changing the B would be a big step.

Protection?

The P-word conjures up images of looking after individual birds, as did the RSPB at its inception. It is a word more suited to an animal welfare organisation than a nature conservation one. It sounds, to many of us, rather too conservative, as though it is more about keeping things the same than the dynamic business that is modern nature conservation.

Society?

This word concentrates on the membership aspect of the organisation rather than its mission. It conjures up a group of like-minded people with similar views, perhaps getting together for their enjoyment rather than to save wildlife. Newer conservation organisations such as Friends of the Earth, Greenpeace, Plantlife and Buglife chose not to mention their supporters in their name.

Royal?

If you had the choice today of your organisation being associated with the Royal Family, or not, would you take it?

You have to wonder to what extent the Royal Charter helps the RSPB do its conservation work and to what extent it hinders. I wonder what the RSPB membership would think on this issue alone? If put to the vote would the RSPB's million members vote to keep the 'R' or ditch it? I, for one, would ditch it.

Once you ditch the 'R' then all the other letters come under scrutiny too. Would a change of name to BirdLife UK really be a loss or would it enable a new beginning? Or even, to mark the broader focus of the organisation these days, Nature UK? I expect devising great new names to become a popular pastime at birding pubs and genteel dinner parties everywhere!

Conclusion

All organisations face choices and challenges all the time. The RSPB evolved enormously during the time that I worked for it and it's difficult to imagine that it won't evolve even more in future.

An organisation dependent on public support suffers tension between doing the best possible job for its mission – saving birds and other wildlife, in the case of the RSPB – and garnering public support (without which no action can be taken). I've always thought that the RSPB should do more to canvass the opinions of its members concerning the organisation's performance (I'm a member and I don't remember being asked) but how do you ask all those potential members, who might be the route to an even bigger and better RSPB, what they think? Might a currently conservative membership actually hold the RSPB back from taking things to the next level?

I offer no solutions to these conundrums here – I am just one in a million RSPB members now and gave up the chance to influence these matters when I gave up my job. But I do think, and I might not have thought this while still working for the organisation, that we members ought to try to have more of a say in what the RSPB does and where it goes. We are the main source of funding for the Society's work and it is done with our money.

Management by objectives works if you first think through your objectives. Ninety percent of the time you haven't.
Peter Drucker

Chapter 16

The tangled bank

It is interesting to contemplate a tangled bank, clothed with many plants of many kinds, with birds singing on the bushes, with various insects flitting about, and with worms crawling through the damp earth, and to reflect that these elaborately constructed forms, so different from each other, and dependent upon each other in so complex a manner, have all been produced by laws acting around us. These laws, taken in the largest sense, being Growth with reproduction; Inheritance which is almost implied by reproduction; Variability from the indirect and direct action of the conditions of life, and from use and disuse; a Ratio of Increase so high as to lead to a Struggle for Life, and as a consequence to Natural Selection, entailing Divergence of Character and the Extinction of less improved forms. Thus, from the war of nature, from famine and death, the most exalted object which we are capable of conceiving, namely, the production of the higher animals, directly follows. There is grandeur in this view of life, with its several powers, having been originally breathed by the Creator into a few forms or into one; and that, whilst this planet has gone circling on according to the fixed law of gravity, from so simple a beginning endless forms most beautiful and most wonderful have been, and are being evolved.

Charles Darwin

Just as Charles Darwin's tangled bank of organisms is the product of millions of years of evolution by natural selection, so is the tangled bank of environmental organisations the product of over a hundred years of cultural evolution. It seems that we naturally tend to congregate in like-minded groups to pursue our interests and dreams. Is the tangled bank of nature conservation organisations a mix of predators and prey? Is there symbiosis and parasitism? Are there worms that crawl and birds that sing?

The taxonomy and nomenclature of the organisations that deal with nature conservation is complex – and has fewer underlying principles behind it than does the bewildering array of species in a tangled bank. There are the great and the small, the statutory and the non-governmental, the rich and the poor, the land-owning and the campaigning, the international and the local, the old and the new, the good and the average, the ailing and the healthy, the quick and the slow, and the quick and the dead. All human life is here, but is there grandeur in this segment of human life?

As I look back on the organisations with whom I collaborated or argued, or whom I was trying to influence or neutralise during my years at the RSPB, I see a somewhat bewildering array. I will mostly concentrate here on the non-governmental organisations, for at least their ways of working and future fate are to a large extent in the hands of ordinary people like you and me, rather than in the gift of governments. How should NGOs work together or compete, and does nature have the NGOs it needs and deserves?

Who's who?

Government departments

There are different governments in different parts of the UK, and so there are different arrangements of government departments. What's more, they don't remain the same within any one administration. The natural environment and nature conservation have had their Westminster home in departments variously called the Department of the Environment, the Department for Environment, Transport and the Regions, and currently the Department for Environment, Food and Rural Affairs (Defra). So the policy subject companions of wildlife conservation have been regional policy, transport policy, planning policy, climate change policy, forestry policy, fisheries policy, and farming policy, at various times. Whose gang you are in, and who else is in the same gang, makes a difference as to how you are treated within your home department.

It must be possible that as the current round of cuts deepens, the number of departments will shrink and that the coalition government will tinker with the departmental landscape – although to date, and probably to its credit, it has not. Defra is a small department and the Department of Energy and Climate Change (DECC) is small too – a re-merger of climate change and environment must be a possibility, perhaps with agriculture being moved to the Business Department where it would become a small fish in a very big pond.

One of the consequences of the creation of Defra, following the foot and mouth outbreak of 2001, was that wildlife ended up in a department with agriculture and fisheries. This was good from the point of view of getting some joined up environmental thinking, but it also put wildlife at the mercy, it seemed, of a large number of agriculture-orientated civil servants, some of whom had little natural sympathy for nature.

Watchdogs and others

Lest we forget, most of the recent extinctions of bodies involved with the natural world have been where quangos (quasi autonomous non-governmental organisations) were killed off by the current coalition government. Bodies such as the Sustainable Development Commission and the Royal Commission on Environmental Pollution, whose opinions and written reports have been important in influencing environmental thinking in the UK over the years, are no more. Their demise leaves a gap where solid intellectual analysis and independent commentary on the state of the environment once flourished.

Statutory sector

Natural England (NE) is the statutory body most concerned with nature conservation on the ground in England. Its counterparts elsewhere are Scottish Natural Heritage (SNH), the Countryside Council for Wales (CCW) and the Northern Ireland Environment Agency (NIEA). All these organisations are different and have different responsibilities and origins. NE, SNH and CCW all had part of their roots in the UK Nature Conservancy Council, but post-devolution they have each gone their separate ways and been merged and coupled with other interests in different ways.

In England (and Wales), there is also the Environment Agency (EA), which deals with flooding and the water environment (and a host of other issues, such as radioactivity), and the Forestry Commission (FC), which is technically a non-ministerial government department but comes under Defra's ambit and is, in many ways, just like another statutory agency.

In England, NE, EA and FC have been told, very firmly, that they are not agents of policy any more. Nor are they now allowed to make public pronouncements except of the dullest sort, as their role is merely to deliver government policy. They are delivery agents and must keep quiet and keep their heads down. In some ways this is fair enough, and delivery is a big job, but again it leaves a gap in terms of elaborating and pronouncing on policy regarding the state of the environment – a gap which NGOs might naturally fill.

Gone are the days when the NCC could launch a document describing the damage caused to wildlife interests by large-scale afforestation of the Flow Country, as they did in 1987, or when English Nature could point out the dangers of herbicide-tolerant GM crops to wildlife, and call for large-scale assessments of their impacts. We cannot expect that sort of thing to happen again in the near future, whatever the pressures on the natural environment. Natural England may be saying behind closed doors that the habitats regulations are not a brake on the UK economy but we can no longer expect to hear them say it out loud – at just the time when it needs to be said.

The NGOs

If you thought that the brief description of the government and statutory sector given above was confusing and complex, then I sympathise. But that's how it is, and it is straightforward compared with the NGO sector. Anyone can set up an NGO and it sometimes feels as though most people have – I expect there will be some more coming along in just a moment!

The really big players in this world are the RSPB and the Wildlife Trusts. Their *raison d'être* is nature conservation – that's their business and they each have huge public support for their work. But, they go about things in different ways.

The Wildlife Trusts are a group of separate, local(-ish) organisations that operate, some better than others, under separate brands or banners. Collectively they manage a lot of land, collectively they have a lot of members, and collectively they do a lot of good for nature conservation on the ground – and their focus is the whole of nature, not just one feathery or fluffy part of it. However, they do not always act collectively and the variation between the best and worst county wildlife trusts is enormous – a bit like the difference between your local corner shop and your not so local Tesco megastore.

The Wildlife Trusts' ability to generate collective policy is weak and their advocacy is patchy because they only have a few staff dedicated to these roles. On a good day (and they have many) and on a good subject (particularly marine conservation and planning recently) they are fantastic. On a bad day (everyone has them), and on a bad subject (the role of Natural England, the future of SSSIs and state woodlands), they can seem like a menace to other nature conservation bodies, and to undermine the conservation of nature.

The RSPB is almost perfect (say I!) and its strengths lie in a very strong analysis of what is affecting birds and the ability to deploy solutions to solve those problems, employing advocacy, a strong scientific capacity and a network of carefully chosen and generally well-managed nature reserves. The RSPB's weaknesses are perhaps that it has a smaller footprint on the ground than the Wildlife Trusts, that it is too birdy, that it doesn't really deal with all of nature, and that it spreads itself very widely – which has to mean thinly, including, these days, drifting off to save birds around the world.

It may have surprised some readers that I didn't count the National Trust among the big guys – it is bigger than the RSPB, the Wildlife Trusts and quite a few other wildlife NGOs put together in terms of membership numbers, and it manages large areas of land. But the trouble is, its focus on nature conservation is weak and blurred. The National Trust dips into nature conservation when it suits it, and is clearly a force for good rather than evil, but saving nature is not its core business and so, quite frankly, it grossly underachieves in this regard. Over the years the National Trust has become less and less focussed on nature and more and more focussed on being part of the entertainment industry – a very good part of the entertainment industry, but nonetheless its mission is apparently to offer you tea cakes in nice surroundings rather than to change the world. If you think that's an unfair assessment then visit the Trust's website, put wildlife into the search engine and see what you come up with.

WWF UK comes and goes as far as UK nature is concerned. Their chief focus is on sustainable development and climate change, but in the marine environment they still have a lot to offer. If only they were prepared to re-engage with wildlife issues then they could make a big impact.

The Woodland Trust is a large landowner but not a large player in nature conservation. Its focus on tree cover *per se*, rather than the nature that might live in those woods and forests, leads to disappointingly simplistic views about nature conservation. It has consistently underplayed the importance of woodland management to the extent that its lobbying has sometimes been to the detriment of woodland nature.

Friends of the Earth were once a powerful force for good in UK nature conservation. They played an important role in getting better protection and management for wildlife sites but they have left the scene to concentrate, it seems, on climate change to the exclusion of everything else. It is difficult to know whether FoE could be effective in helping nature in the UK in future, but it is clearly largely irrelevant to it at the moment. I, for one, would welcome them back wholeheartedly as their brand of outspoken challenge would spice things up a bit.

Greenpeace have never been that involved in UK nature conservation and probably never will be – but they are a well run and effective organisation who, again, could make a real difference if they chose to play a part.

The Marine Conservation Society does a good job in standing up for marine issues and helps to make sure that they are not neglected by other NGOs or by government.

Butterfly Conservation is younger than I am, and Plantlife and Buglife are younger still, even when their ages are added together. These three forty-something, twenty-something and almost teenage organisations are relative newcomers which cannot be judged by the standards of their much older and larger fellows. Each is, however, focussed on their particular part of nature. Isn't it remarkable that all those plants and all those insects haven't gathered more support than those few birds?

Let us not forget the Wildfowl and Wetlands Trust with its surprisingly large membership of over a quarter of a million, nine fine reserves and visitor centres dotted across the UK, and an increasing involvement in ambitious conservation projects and advocacy. If it can recapture some of the conservation ethos of its founder, Sir Peter Scott, then it could become a much bigger player on the UK conservation scene.

The British Trust for Ornithology, whilst a fast-developing organisation that is clearly going places, does not fit in this list. It is an organisation for the caring birder but it is not a conservation organisation – it dips its toes into nature conservation now and again but whenever the water gets a bit hot it will be nowhere to be seen, and that will limit its popular appeal into the future. This is a policy decision, having decided to put providing impartial information firmly ahead of advocacy. What the BTO does, collecting information and coordinating birders in a fantastically effective way (why doesn't this happen for plants or insects, or mammals?) is very useful to nature conservation, but it isn't nature conservation.

There are a variety of other nature conservation organisations, including Amphibian and Reptile Conservation, Badger Trust, Bat Conservation Trust, Campaign Whale, Grasslands Trust, Hawk and Owl Trust, Mammal Society, Marinelife, National Trust for Scotland, Pond Conservation, Shark Trust and the Whale and Dolphin Conservation Society – all of which are members of the Wildlife and Countryside Link (London version – other varieties exist in other centres) and which therefore play a part in the development of joint NGO policy and advocacy.

Everyone else

Of course, nature conservation is just one issue, with a variety of organisations working to further its aims. There are lots of other issues too, from field sports to agriculture, and from renewable energy to walking in the countryside, all of which may have their own NGOs and statutory agencies, as well as industries, associated with them, who all keep an eye on what's going on in nature conservation NGOs and what they are doing and saying.

The last decade has seen the rise of the concept of the 'stakeholder'. It seems that everyone is a stakeholder now and it is the job of every government department to attend to, foster, or contain its stakeholders' interests.

Working together

When Charles Darwin contemplated the variety of species on Earth, their various forms and the apparent perfection of their fit to the world in which they lived, he was looking for an explanation for this current reality. How had we got to this position? His theory of evolution by natural selection provided a sweeping explanation for that variety of life.

In contemplating the variety of nature conservation organisations, I think we should be less interested in how we got here and more interested in where we go from here. In particular we should think about whether, if we put ourselves in the position of the threatened wildlife of the UK, we would find the current disposition of wildlife NGOs a perfect match to our needs? Is nature well served by UK NGO nature conservation? Is there grandeur in this scheme of things?

There must be an awful lot of threatened wildlife species who wish they were birds! The RSPB is not by any means perfect but it has an unrivalled record of turning around the fate of declining bird species such as the bittern, corncrake, red kite, stone curlew, cirl bunting, avocet, marsh harrier, osprey amongst many others. We could legitimately say that the RSPB didn't do all this alone, others were certainly involved, but you need someone to drive conservation action forward and in all these cases, without the RSPB, these species would be far worse off. What is the list of conservation successes achieved by other wildlife conservation organisations? Who else has turned around the fate of species which had been in UK decline, or were nationally extinct, and put them back on the road to recovery? There are some examples. The work of Butterfly Conservation on several butterfly species, notably the large blue with which Jeremy Thomas is heavily involved, stands as a good non-avian case. But I don't think there are that many other such examples out there.

And this does make you think about whether every taxon needs an RSPB. Should there be an RSPButterflys, an RSPPlants and an RSPMammals, in the sense that the threatened wildlife of different groups needs a large and technically proficient wildlife NGO to look after its interests? This might be the thinking, and I believe in some sense it has been, behind the emergence of a number of the newer wildlife NGOs. But none has been able to approach the size and power of the RSPB, and to be frank, it doesn't look as though any of them will in the next decade or so. That means that threatened plants, insects etc. may go to the wall without a powerful ally to save them.

Plantlife, one of my favourite nature conservation organisations, is too small to do a job for all of the plant species in the UK – lichens, mosses, ferns, trees and flowering plants – remotely similar to that done by the RSPB for birds. And that won't change very quickly unless Plantlife finds some very rich benefactors to bankroll its work. It is a good plant conservation organisation at the moment but it cannot meet the needs of threatened UK plants without being something like a hundred times bigger – an RSPPlants in effect. I can't see it happening although I would be delighted, truly delighted, if it did. There aren't many different sources of money available and the facts of life are that the public and sponsors tend to reward success, which means that it is difficult to get a new organisation off the ground.

Imagine if Plantlife were the size of the RSPB and the RSPB were a small, new, bird conservation organisation with a modest number of successes under its belt. Would it be easy for the RSPB to grow today? I think not. Many of the newer nature conservation organisations seem to hit a ceiling at around 10,000 members. I'm not sure why, but it must reflect the difficulty of finding funding and perhaps means that without a track record of success it is difficult to persuade the public and others to give you more resources, while without the resources it is difficult to generate the success – a cruel bind.

Perhaps we should be advocating a mass emigration from RSPB membership to swell the ranks of the other organisations and even up the conservation chances of the plants and invertebrates? Should we aim to transfer a hundred thousand members from the RSPB to each of Plantlife, Butterfly Conservation, Buglife and one or two other excellent nature conservation organisations whose focus is non-avian? Of course not – it should not be either/or. So if you are thinking about which other conservation organisation you should join as well as the RSPB, then these are the top of the list that I would suggest!

If we cannot easily grow RSPBs for other threatened taxa, and it seems that we can't because people are trying with only modest success, then can we change the RSPB into an all-taxa nature conservation organisation? Can, and should, the RSPB shake off that bird thing and become the Royal Society for the Protection of Biodiversity?

If you are an observer of wildlife NGOs as well as an observer of wildlife then you will have noticed that the RSPB changed its strapline from, 'For Birds, For People, For Ever' to, 'A million voices for nature' a few years ago. Pick up the latest RSPB Annual Review and you will read on the back cover that, 'The RSPB speaks out for birds and wildlife' and you are asked to 'help us win even more victories for wildlife this year'. Inside that back cover you are asked to 'step up for nature' because, 'nature is in big trouble' and told that if we act together we will 'help the UK's wildlife thrive once again'. Inside that Annual Review you will find photographs of black grouse, corncrake and red kite, but also of butterflies, hares, lizards and seals. The RSPB wants you to think that it is a nature conservation organisation these days and not just a bird conservation organisation - and it is getting there.

This process has been progressive over a good many years – part deliberate and part unconscious drift. The RSPB has increasingly realised that the tools of the trade it deploys to benefit birds, and that have been described in this book, are pretty much the same tools that need to be deployed for all nature. Site protection, species protection, reintroductions, lobbying government on land use policies, and nature reserve acquisition and management can save the corncrake and the skylark, but also the medicinal leech and the harbour porpoise if used correctly. True, you need to learn a bit about the intricacies of the ecology of a lot more species, but you have the tools at your fingertips so why not deploy them for the greater wildlife good? This is a view that I espoused and encouraged within the RSPB for many years.

I think that if you put yourself in the place of threatened UK wildlife then you would be crying out for the RSPB to spread its expertise and care a lot more widely – there is an undeniable need and it isn't being met at the moment. The need is so large that you can hardly be accused of stepping on someone else's toes by lending a hand in their patch. Except that is exactly the danger, and will be one of two main tensions holding the RSPB back from becoming the RSPBiodiv. What will our friends say?

Would a more open and publicly announced move by the RSPB into 'all areas wildlife' be greeted with unbridled enthusiasm by other nature conservation organisations? Probably not, because they will not be thinking of nature, they will be thinking of their own corporate interests. The world is a competitive place and the family of wildlife conservation organisations is just that – a typical family with its tensions and jealousies, remembrances of past slights (imagined and real), alliances and feuds. The appearance of a tightly-knit and loving group must be upheld to the outside world, although occasionally the mask will slip, but in private there is a fair degree of tension.

There are real differences in approach and philosophy between different wildlife NGOs, many of which are rarely examined, and certainly not in public. The real differences between wildlife NGOs are not whether they 'do' birds or butterflies, all nature or some nature, terrestrial or marine, but whether they promote people or wildlife, whether they talk of species or habitats, whether they see habitat management as vital or optional, whether they are government's friends or foes, whether they see industry as an important ally in saving wildlife or not, whether they take money from anyone or are they picky? These are differences in world view that are based on principle not biology. These differences rarely surface in public.

I think it is interesting that wildlife NGOs don't specifically set out their philosophies for the public to see, and, even more critically, they don't even do a very good job of publicising their nature conservation achievements. They rely largely on whether you are interested in their brand of nature, or their region of operation, rather than proving to you that they are doing a good job.

So the first reason that the RSPB would be nervous about suddenly announcing that it was the RSPBiodiv would be what the other members of the family would think and say. How would organisations such as Plantlife, Butterfly Conservation and Buglife, who largely share the RSPB philosophy on how to do nature conservation, react? No-one wants to fall out with their closest friends, after all. And how would the Wildlife Trusts and others react? You can see that it is a tricky area. But at the moment it looks as though the

RSPB is heading in that direction by stealth and it's not an easy thing to hide. So why not just come out and admit where one is heading – if that is indeed where one is heading?

I said there are two main reasons why the RSPB might be nervous about becoming the RSPBiodiv more openly, and the other is the danger of 'losing' birds. There is no doubt that birds have an appeal that is universal and very strong. Look at Springwatch or Autumnwatch and see how many of the images are of birds when the BBC has all of nature to choose from. How much of the RSPB's support is so dependent on birds that any change would jeopardise that support?

This question needs careful consideration but must be balanced by that of how many more supporters might be enlisted with an avowedly all-nature proposition. I am a birder first and foremost, but I have a moderately good knowledge of many other groups of life on Earth and a warm feeling towards all of them. I know from my professional background that the RSPB does an awful lot of good for non-avian wildlife and I am pleased that that is the case. It salves my conscience on the subject, but, personally, I would be even happier if the RSPB told me that it intended to spend a higher proportion of my membership fees on things other than birds – I might even give more money to them as a result. But the question for the RSPB is how many people are there like me? What would the overall financial impact be?

But let's get back to nature (always a good motto) and what wildlife needs. UK wildlife needs more resources devoted to it, more or stronger voices shouting for its salvation, more influence exerted on government and more friends, more influence and more money. UK wildlife needs that now, not in 50 years time. So if you put yourself in the place of UK threatened wildlife, and then look at the current state of the tangled bank of UK wildlife NGOs, then I think that you would want the RSPB to move into an all-wildlife mode as quickly and as effectively as possible. I think you would want the public to support it, and other nature conservation organisations to welcome it.

If you had been UK wildlife scanning the scene 25 years ago then I don't think that you would have necessarily come to the same conclusion. The RSPB was very birdy then and was only a little way along the road of developing policy and advocacy skills – and its membership and resources were much smaller. At that time you might have suggested that the right 'big move' within nature conservation was that the RSPB and the Wildlife Trusts should merge and work more closely together. Through pooling resources surely you could develop a super-NGO with local strength and national clout, with a strong science base and a network of nature reserves everywhere across the UK.

Around that time, and for those reasons and others, such thoughts were mooted, discussed and then rejected. I guess, although I don't really know, that it was feared that the overlap in membership would mean that a merger would result in a net loss of members. I guess too that among the disparate array of county Wildlife Trusts some felt that they had much to gain and others that they had much to lose from such a proposition. Not knowing the details I can only speculate, but looking back it does look like an opportunity lost. Who knows how it might have worked out but if all had gone reasonably well – and why shouldn't it have – then the UK might have a single, powerful voice for all of nature right now.

I am not suggesting that a merger between the RSPB and the Wildlife Trusts would be a good idea now. I think the moment was missed and the time is not right now. There would still be a variety of views on the prospect from the individual Wildlife Trusts but I think that during the passage of time the two organisations, or the 49 organisations, have grown apart. The RSPB has developed a significant body of international work and the Wildlife Trusts are domestic and talk increasingly of people and less about nature. The differences in philosophy and approach have created a gulf between the two and there may not be quite enough goodwill to bridge that gap. It could have been great if done then – but it almost certainly wouldn't be so great if attempted now. That's what seizing the moment is all about.

The Wildlife Trusts and the RSPB are the two big beasts in the tangled bank of nature conservation NGOs. If they have missed the chance of merging to become a super-beast then, as night follows day, they will be forced to compete for resources and profile – as indeed, they already are. As are all the other wildlife NGOs on the tangled bank.

I am a member of the following organisations: National Trust, Wildlife Trusts, RSPB, Butterfly Conservation, Plantlife, Buglife, British Dragonfly Society, Wildfowl and Wetlands Trust, Pond Conservation and the BTO (and I have probably forgotten someone and there are others from whom I've come and gone over the years). These annual subscriptions come to about £250 and that's a lot of money. Many people I know have similar portfolios of memberships – some driven by financial calculation (how often do I want to park somewhere where I'd have to pay the NT if I weren't already a member) and others by altruism towards threatened wildlife. Some are driven too by the opportunities that the organisation offers to meet other like-minded people through members' events or to make a personal contribution (and also meet others) through volunteering. I would happily join another half a dozen organisations if I had the spare cash.

However, there is no doubt that the numerous magazines that I am sent, reminders of membership renewal, and newsletters etc. are all overheads that my memberships pay for. There is duplication in this tangled bank for the human contributors and benefactors which must detract from the sum of the benefits achieved.

Are there ways, short of merger, where wildlife NGOs could cooperate more to do a better job for the natural environment? I think there are at least four worth exploring.

Kids' stuff

Most wildlife NGOs have children's membership and activities and yet these children rarely recruit immediately into the adult membership – the hope is that they will eventually and that, even if they don't, they will be better-informed adults as a result of a bit of pond-dipping when young. I would like to see the wildlife NGOs all getting together to set up a single 'kids for nature' and 'nature for kids' organisation that dealt with children properly with a national network of events and activities. Such a thing would be an attractive sponsorship opportunity for industry and grant-giving bodies alike (for some reason everybody likes children).

If we really believe in enthusing young people then let's do it properly and do it for all kids and all nature.

Boring backroom stuff

Every NGO has computers, staff, members, and finance departments and all these need to work efficiently, and all cost money which could otherwise be spent on saving wildlife. The large and the small need to upgrade their computers every now and again and to learn new health and safety legislation, and all need to have a membership system. This leads to duplication of effort which would disappear if a few organisations merged. But even without merger savings could be had through sharing resources and expertise, and this could be done on a purely commercial basis. I'd like to see the larger organisations offer to carry out some of these tasks for the smaller organisations at a rate that would save money. I think it highly likely that the National Trust could offer Plantlife a membership service that would be much cheaper than Plantlife could do themselves, without me, as a member of Plantlife, noticing any difference at all. Perhaps, the smaller wildlife NGOs should get together and invite the larger ones to bid for a contract in order to get the best possible deal?

Nature reserve management

A similar philosophy to the backroom service could be developed for nature reserves within a region. Each organisation has its own wardening staff but also a whole lot of expensive machinery too. The cost of land management is a high proportion of the overall expenditure on reserves for any organisation that takes it seriously, so pooling machinery and manpower resources could be very valuable. It happens to a small extent now. But all would benefit from these practices developing further.

Advocacy

This is the area where most collaboration already occurs. I have sometimes gone into a meeting with a government Minister on my own, representing the RSPB, but rather more often I have been part of a group with agreed aims and a prepared script. NGOs are good at presenting a pretty concerted view to outside organisations.

But, there is scope for NGOs to work together even more effectively to influence decision makers. If you look at the most marginal parliamentary seats there are a great many where the total membership of wildlife NGOs could influence who wins that seat in the next general election, and therefore the shape of the government. I am sure, and this is an idea that has been knocking around for a while, that there is enough time before the next general election for wildlife NGOs to warn the political parties that they intend to shine a light on the parties' environmental policies in these seats at election time.

Conclusion

Maybe you think that my analysis is clouded by long years of working for the RSPB – it probably is, and it would be odd if it weren't. One thing I would say is that I can't recall ever applying for a job with another wildlife NGO, despite occasionally being tapped up by them. I always looked at what the RSPB was doing and thought I would rather stay where I was to make the biggest impact I could on nature conservation. The RSPB starts from an analysis of what nature needs and then tries to provide it – and that seems a good place to start and a good thing to do.

This approach, giving nature what it needs through science, practical conservation work and advocacy, is a tried and proven route for conserving wildlife. It will and does work for every type of wildlife species and there are many threatened species out there that would benefit from it – and rather few of them are birds.

And so, if I were given £100 and told to distribute it to UK nature conservation organisations I would give it all to the RSPB as that's where I think it would do the most good. If, on the other hand I were given £1m to distribute then I'd probably split it equally between Butterfly Conservation, Buglife and Plantlife but I'd ask a few probing questions about what they might do with it. I would want to see it spent on putting them on a more sustainable route to growth so that they could conserve more neglected species. If I had £10m at my disposal than I would go back to the RSPB with the following proposition: 'I'll give you £1m today, because I think you are wonderful. I'll give you another £1m in six months time if you provide me with a plan, signed off by your Council, for the RSPB to increase the effectiveness of its conservation of non-avian wildlife. You'll get another £2m if I like your plan. After that I'll give you up to £1m a year for the next six years as you put the plan into place, depending on how well I think you are doing. Any money that you don't get will be split equally between Buglife, Butterfly Conservation and Plantlife'. That might get things moving!

The NGO world is a little half-hearted; there is competition but it is very gentlemanly, and there is collaboration but it doesn't go very deep. If you could poll the starfish and the bluebells, the harvest mice and the chanterelles, then they might ask for a better attempt by us to save them. I think they might point to the need for fewer organisations and more resources to come to their aid.

Let a hundred flowers bloom, let a hundred schools of thought contend.
Mao Tse-Tung

Chapter 17

What we need to do to win

This is a beautiful planet and not at all fragile. Earth can withstand significant volcanic eruptions, tectonic cataclysms, and ice ages. But this canny, intelligent, prolific, and extremely self-centered human creature has proven himself capable of more destruction of life than Mother Nature herself… We've got to be stopped.
Michael L. Fischer

Imagine the world before humankind had made much impact on it. You don't have to go back very far – 20,000 years will do it. The seas were full of fish and whales. Tropical rainforests girdled the centre of this blue planet. Giant ground sloths lived in north and south America and Europe had Irish elk, woolly mammoths, woolly rhinoceroses, and cave hyenas too. With the spread of humans from Africa and Asia much of this megafauna was lost – partly due to hunting but also perhaps due to natural climate change and the spread of diseases. It looks like our ancestors' prowess with the spear was responsible for many of these extinctions as the species that vanished were mostly those that we could eat or that might eat us – whereas many smaller mammals were apparently unaffected by whatever was going on about 10,000 years ago.

You could say it has been downhill for all the other species on the planet ever since, with humankind (human-unkind more like!) spreading in range, increasing in numbers and having an overwhelming impact on every continent and ocean, and now even on the atmosphere itself. We are watching the sixth global extinction crisis, but with species disappearing from the planet at a rate between 100 and 1,000 times the 'natural' rate.

I sometimes wonder what a visitor from outer space would make of it. If their species had made occasional visits to Earth every 10,000 years or so they may have had it in the guidebooks as a lovely place – until their most recent trip when the updated entry might have read 'going downhill'.

Yet our own species has done very well by many measures. Our numbers have increased hugely – we are not a conservation case. Ten thousand years ago there were probably only about five million of us and the human population reached one billion only at the very beginning of the 19th century. My father was born in 1920 when the world still hadn't quite accommodated its two billionth standing, living human, but my birth, hours before the 1958 Grand National, helped bring our species to three billion by 1959. We had clocked up five billion before my daughter and son were born in the late 1980s and early 1990s, we hit six billion in 1999 and seven billion while I was writing this book.

We have done pretty well over those last 10,000 years. You and I don't live in caves and we don't fear being eaten by strong, fierce mammals such as sabre-toothed tigers. We have done many marvellous things – we have invented football and produced exponents of it such as Pele and George Best; we've written novels such as *Far from the Madding Crowd* and *the Hitchhiker's Guide to the Galaxy*; and produced music from *Tosca* to *I Say a Little Prayer for You*. We've invented mobile phones and the internet and democracy and religion.

We've also realised that we aren't at the centre of things – we are on the third rock from a rather average sun in a very big universe, and our species wasn't put here in all its fully formed perfection but evolved alongside the rest of life on this amazing planet. And we have recently begun to understand how this complex place works – how carbon cycles through ecosystems and how ecosystems are interconnected so that tinkering in one place can lead to changes in another.

We call all this 'progress' and until now it has seemed that every generation of people on Earth would have a safer, richer, easier, more pleasant life than the last. Certainly my life has been easier than my father's, and his easier than my grandfather's, but I don't think that that will necessarily be true for my children. That's not just because of some enormous failure of the world's financial system, which no-one seems to understand, but is due to the ecology of the Earth. We have got to the point where the way we live, and the number of us trying to, prejudices this so-called progress as well as having a massive impact on the other species around us. Just as crashing numbers of farmland birds don't bode well for the sustainability of our food production, so the accelerating extinction of species should send us all a warning about where we are heading.

We baby-boomers may well be the most fortunate generation ever – we have enjoyed the peak of the good times and will leave it to our children and grandchildren to try to sort out the mess which previous generations have created. It certainly wasn't all our fault but it would be good if my generation could say that we helped to solve the crisis of how to live on this planet sustainably.

We need to get on with it. All the indications are that population growth and unsustainable lifestyles are going to make this planet a far nastier place for many of our fellow humans, including our children, than it has been for us. It feels like we are driving towards a brick wall and instead of reaching for the brake we are pushing the accelerator. If only we had enjoyed our current understanding of the world, and how it works, in the year 1900 or 1800, rather than 2000. How could we have spent that century or two in getting things right? Would we now be in the mess that we are in or would we now have found the brake, avoided the wall and be living a life of luxury with only three billion of us on the planet as a result? Maybe this is the biggest case of underinvestment in science that the world has ever seen? Maybe we'd have screwed it up anyway?

This book started with me telling you that my Dad used to point out birds to me in the countryside where we lived. That interest led, through a series of accidents, to me becoming a scientist studying animal behaviour and then a conservationist trying to stop some of the losses of our wildlife. If you spend your life, as I have spent more than 25 years of mine, trying to make the world a better place for wildlife, you come to realise that often a better world for wildlife is a better world for people too. The ultimate causes of species loss, whether they are in the Amazon Basin or the Nene Valley, are over-exploitation, habitat destruction, exotic species and pollution. None of these things are good for our species' future either.

I think it is almost inevitable that a strong interest in wildlife is likely to make someone an environmentalist, or at least more of an environmentalist than they would otherwise have been. It has done that to me and I have seen the same happen to so many others too. Whether you are conserving bitterns or skylarks or whales or orchids you will always trace their problems back to man and usually some unintended consequence of the way we live our lives. There are few counterparts of our ancestors' habit of driving mammoths to extinction by killing them with spears – though remember Japanese whalers, and gamekeepers waging war on hen harriers. Instead, these days, most of the problems facing wildlife are indirect ones linked to the way we live. Farmers with spears or guns haven't dramatically reduced the number of skylarks in our countryside, but farmers, pesticide companies, consumers, supermarkets, civil servants and politicians have. We're all in this together except that the victims, our wildlife, don't get a say in things unless we speak up for them. Only some farmers and a very few pesticide companies or supermarkets will take account of the skylark's needs and that is why there are so many fewer of them. Politicians and civil servants, and in the end supermarkets, pesticide companies and farmers, will only take notice of the needs of the skylark if we make them, for we get the politicians, civil servants, pesticide companies and supermarkets we deserve!

What we need, and it's a vanishingly long shot, is an environmental messiah figure. Someone whose message to the world is so convincing and compelling that billions of people will take notice and governments change course to address the ecological threat to us all, to the world and the millions of other species we share with it. Perhaps a US President or, perhaps even better, a Chinese or Indian leader. Or perhaps just a peasant, or a street cleaner, or a primary school teacher, who catches the public imagination on TV one evening – and is free from any vested interests or political or religious baggage? A person who would take the scientific position on why we need to change the way we live and link it to a moral position showing how unfair it is to future human generations and current wildlife generations to trash the planet – particularly when we don't have to. We can live rich and fulfilling lives and yet drastically reduce our ecological footprints with a bit of restraint and thought.

But in the current absence of a charismatic messiah, we'd better get on with it ourselves. You and I are not without responsibility for how the world is but then neither are we without power and influence. My favourite quotation of all time is that attributed to the late 18th century Whig statesmen, author and philosopher Edmund Burke: 'The only thing necessary for the triumph of evil is that good people do nothing.' Now, there is little evidence that Burke ever actually said this or wrote it, and if he had, then he would, at the time, almost certainly have said 'men' rather than 'people' but nevertheless it is a credo to live by.

You and I do not have to be victims – we can choose to make the world a better place. And if we choose not to get involved then we are actually choosing to let the world go to pot. So here is a bullet-point manifesto for action for the motivated, active naturalist, birder or environmentalist:

- Revel in the natural beauty of the world. You and I are lucky enough to be alive, to be reasonably well off, and to live on the only place in the universe where we can see a blue whale, a holly blue butterfly, hear a blackbird sing or smell pine trees. It is the natural world that makes this planet so special and it's ours to admire and enjoy – it lifts us up from the daily grind to a higher plane.
- Study and research the natural world and environmental issues. Read books (thank you for reading this one!), search the internet and talk to people so that you gain understanding and knowledge. Don't take anyone's opinion as your own – make up your own mind on the evidence.
- Share your love of the natural world and your knowledge of it with others. You don't have to be an expert to be an enthusiast and you don't have to know everything to know more than the guy you meet on the train or in the pub. Invest in children's love of nature because they are the future generation – but don't neglect the adults – they should be better informed and you can play a part in that too.
- Join forces with like-minded people. Join environment and nature conservation groups – give them your financial and moral support and volunteer if you can do a little extra to help them.
- Spend your money to influence change. Buy wisely, though it's often difficult to know exactly what that will mean. Tell the people you buy from that you are supporting them because they are environmentally better than their competitors whether it is because of recycling, organic farming, energy use or whatever. Money talks and you can make your money shout if you put some effort into it.
- Cut your carbon footprint. Work out your current carbon footprint (there are lots of places on the internet to do this) and then see what you can cut out. Set yourself a modest reduction target for the first year, and one that will save you money, and see how you get on. You'll find it easier than you think and you'll want to do more.
- Eat less meat. Start by giving up meat for one day each week and you'll find it is very easy and you will probably want to do more. You won't really notice the difference but the knock-on effects on food supply, climate change, feeding the world and wildlife will be positive ones.
- Get politically active. Know what your MP says and does and let them

know that you care about the environment. Join a political party (any one will do though I have admitted my own preference) and work within it to make it more environmentally aware – we need that environmental messiah to come from somewhere and it might just be from your local party meeting.

- Recognise the enemies of environmental progress. Once you have your own environmental agenda you will recognise that others are working against it. At present my top three might be: the Chancellor, George Osborne; the NFU; and those gamekeepers who shoot hen harriers. Don't let their views go unchallenged and don't let them think that the world hasn't noticed what they are saying, or agrees with it. You must choose your own enemies, but have a look around and you will surely find some.
- Revel in the natural beauty of the world – it's so important that it's worth having in this list twice, and it's where all such lists should begin and end.

> *Modern society will find no solution to the ecological problem unless it takes a serious look at its lifestyle.*
> **Pope John Paul**

Index